Nebraska
STATE HISTORICAL SOCIETY

First Telegraph Line across the Continent

Charles Brown's 1861 Diary

Edited by Dennis N. Mihelich
and James E. Potter

Nebraska State Historical Society Books
Lincoln

Brown's diary is published with the permission of the Smithsonian Institution, Washington, D.C. (Western Union Telegraph Company Records, Archives Center, National Museum of American History, Smithsonian Institution).

Unless otherwise credited, photographs in this book are from the collections of the Nebraska State Historical Society. Archival photos are identified by their Record Group (RG) number. To obtain copies of photos, or to request publication permission, contact NSHS reference staff at (402) 471-8922, or email nshs.libarch@nebraska.gov, or visit www.nebraskahistory.org

Book design by Reigert Graphics, Lincoln, Nebraska

ISBN 978-0-933307-32-2
Library of Congress Control Number: 2011927234
Printed in the U.S.A.

Publication of this book was made possible by
The Ronald K. and Judith M. Stolz Parks Publishing Fund
established at the
Nebraska State Historical Society Foundation,
and use of this Fund for this purpose
is made in memory of
Wayne Kemper Parks (1909-1995)
and Hazel Virginia Hill Parks (1911-1991),
lifelong Nebraskans who were born on Madison County farms,
were married on March 19, 1930,
and were farmers in Madison and Pierce counties.

Illustrations

Contents

Prologue

In 1861 Charles H. Brown, a twenty-seven-year-old lawyer, became Edward Creighton's "Man Friday." Both men were recent migrants to the city of Omaha in the newly established territory of Nebraska. Creighton trekked from Ohio, building telegraph lines, hauling freight, and grading city streets and railroad rights of way. Brown followed the route of two older brothers who had migrated to Omaha in 1856 and obtained employment freighting for Creighton before establishing a mercantile outlet. The family connection probably secured Charles Brown a teamster's job with Creighton in 1860; his good work and his education likely ensured his promotion to the position of right-hand man to the superintendent of construction for the transcontinental telegraph. While seeing to Creighton's business correspondence, supervising the off-loading of supplies in Omaha and their distribution to job sites, and participating in blue-collar construction work, Brown had time to log a diary,

"My Experiences on the Plains in 1861 in assisting in the Construction of the first telegraph line across the Continent."

Brown's diary survives as the only known chronicle of the daily events associated with the building of the transcontinental telegraph. In its detailed descriptions, it surpasses the short, business-oriented article "Wiring A Continent: The making of the U.S. Transcontinental Telegraph Line" (originally published in *The Californian* magazine, 1881) by James Gamble, superintendent of construction of the western segment of the line. Brown penned a primary historical account of inestimable value. It is part travelogue, describing the various means of transportation, the local topography, flora, fauna, and climate, and the routine at the forts and road ranches along the way. It is also history from the workers' point of view: what they ate and how they slept, amused themselves, hunted, forded streams, gathered the poles, dug the holes, and strung the wire. The reader becomes a member of the crew, experiencing the arduous but enjoyable and fulfilling work (except for battling hordes of ravenous mosquitoes) of building the eastern half of the transcontinental telegraph.

Moreover, Brown's account presaged Mark Twain's *Roughing It* (1872); both imparted colorful stories of the sparsely settled West (Twain rode a stagecoach from St. Joseph, Missouri, to Sacramento, California, while Brown traversed part of that route, Julesburg, Colorado Territory, to Salt Lake City, Utah Territory, on foot and riding mules, horses, carriages, and stagecoaches). Brown also exuberantly espoused the "continentalism" of John Quincy Adams (with divine guidance Euro-Americans would develop the landmass between the Atlantic and Pacific oceans) and the "Manifest Destiny" of journalist John L. O'Sullivan, editor of the *Democratic Review* (Providence proclaimed the right of Euro-Americans to control the continent). In that vein, the diary presented a literary equivalent of John Gast's painting,

***American Progress*, an 1872 painting by John Gast. This iconic image represents the nineteenth century notion of Manifest Destiny, an idea that is prominent in Charles Brown's diary. Autry National Center; 92.126.1**

American Progress (circa 1872), which depicted Euro-American civilization conquering the West (the telegraph and the railroad occupy the center-right portion of the canvas). As he jotted in his diary, Brown prayed for the preservation of the Union then being fractured by the Civil War (July 4, 1861), advocated the building of a transcontinental railroad (July 10, 1861), and presented plans for the utilization by miners, farmers, and ranchers of virtually every acre of ground he navigated.

About the Native Americans who occupied the Great Plains, Brown maintained conflicting views, while never doubting the racial superiority of Euro-Americans. He considered Indian men lazy and uncivilized, referring to them as "savages" in most instances. Yet he described Old Spot, the chief of a Cheyenne band, as "a very dignified and courteous man, about fifty years of age, fully six feet tall, powerful frame and a good looking intelligent Native American." He compared Old Spot's diplomatic skills to those of Daniel Webster and Henry Clay (June 29, 1861). Furthermore, while Brown decried the rampages of the "savages," he attributed much of the violent conflict resulting from Manifest Destiny to the actions of the white pioneers, who sought to drive the Indians from their ancestral lands. "It is a great wonder that the whites are accorded such lenient treatment as they have from these wild men of the Plains," he wrote on June 19, 1861. Brown also provided significant ethnographic data, detailing the appearance and lifestyles of the Native Americans he met.

Thirty years after his adventure, Brown copied his diary in pen and ink using a cursive style. Between December 28, 1890, and June 25, 1891, he added a twenty-six-page introduction, which established the historical context for building the transcontinental telegraph. It also related amusing and enlightening anecdotes of the movement of laden wagon trains and of stagecoach travel along the Platte River Road between Omaha and Fort Kearny, Nebraska Territory. Inexplicably, pages eleven through fifteen are missing; the total manuscript totals 121 handwritten pages. The diary began on June 18, 1861; the first entry described Brown's departure from Fort Kearny, and the entries for the subsequent eight days detailed his trek to Julesburg, Colorado Territory. He amended the June 20, 1861, entry on January 30-31, 1894, adding an extensive depiction of a band of about fifty Sioux Indians he had encountered. He arrived at the six-building town of Julesburg on June 25, 1861; during the ensuing week he wrote about the local climate, the comings and goings of the stagecoaches, and the preparations for the commencement of telegraph construction.[1]

On July 2, 1861, Brown noted, "We commenced the construction of the telegraph line to-day." He helped Edward Creighton dig the first hole. The declaration annuls the myth that had rapidly taken hold claiming that construction began on July 4. The symbolism of beginning construction on Independence Day of a device that would bind the nation together, when in fact it was being torn asunder by the Civil War, became standard in historical accounts. The remaining entries meticulously explained the process of construction, from gathering the correct-sized trees to use as poles, to walking off and marking the poles' spacing, to digging the holes, erecting the poles, and stringing the wire. Brown concluded his journal on August 9, 1861, with a short four-sentence notation. The day before, he had explained that Edward Creighton had demoted his brother, Joseph Creighton, and put Brown in charge of that wagon train. Seemingly, the new assignment that had Brown working away from camp and ranging across the countryside in search of poles inhibited his nightly jottings. Much work remained before the line reached Salt Lake City. Unfortunately, that part of the story remains lost or untold. However, the narrative Brown did record is historically significant and enjoyable reading.

"What hath God wrought!": The Evolution of Telegraphy

On May 24, 1844, Samuel Finley Breese Morse, an accomplished painter turned scientist and entrepreneur, sent a Biblical phrase (Numbers 23:23) by telegraph from the Supreme Court chamber in the United States Capitol to his partner Alfred Vail sitting forty miles away in Baltimore. Surprisingly, this milestone in long-distance communication failed to ignite the imagination; initially, government officials, military leaders, and businessmen greeted it with apathy. The indifference, however, quickly evaporated as innovative entrepreneurs comprehended the value of, and the demand for, rapid and wide-ranging communication. Within two decades transmission wires connected all major American cities, and a transcontinental telegraph line conquered the vast space between the East and West coasts.

A Pony Express rider saluting the telegraph builders (who would soon put the Pony Express out of business) was a popular motif in illustrations of the period. NSHS RG24090-144

First Telegraph Line across the Continent:
Charles Brown's 1861 Diary

From the dawn of civilization, delivering information between distant places was a slow, arduous, and unsophisticated process. In 490 B.C. as the mythological story goes, Pheidippedes ran twenty-six miles from Marathon to Athens to announce the Athenians' victory over the Persians. The Greeks also used fires from hilltops to convey simple messages: one, two, or three fires each had a specific meaning (e.g., "All's well" or "Send help"). Two thousand years later the English used the same means to warn of the approach of the Spanish Armada in 1588. Signals using flags or lighted torches originated in the seventeenth century.[2] Hollywood movies made legendary the American Indians' use of fire and smoke, as well as romanticizing the exploits of those who rode for the short-lived Pony Express. Moreover, for more than a century every American student learned of *Paul Revere's Ride* from the poem by Henry Wadsworth Longfellow:

Hang a lantern aloft in the belfry arch
Of the North Church tower as a signal light—
One, if by land and two, if by sea;
And I on the opposite shore will be,
Ready to ride and spread the alarm
Through every Middlesex village and farm,
For the country folk to be up and to arm.

Fifteen years after Revere completed his celebrated jaunt, the Frenchman Claude Chappe devised a system that eliminated the services of a rider. He invented a pivoting wooden panel that worked in conjunction with a clock; the arms moved to indicate the number on the clock face. A person on a distant tower, using a telescope, could see the indicator and relay it to another far-off tower. Chappe formulated a code for the numbers to relate to the letters of the alphabet. Thus, optically, a message moved through the air much faster than a courier on a horse could deliver it. He wanted to name his invention the *tachygraphe*, Greek for "fast writer," but a friend, Miot de Melito, a government official and classical scholar, suggested *telegraphe*, meaning "far writer." On July 2, 1793, a communication sent using the Chappe semaphore system garnered the moniker *telegramme*, the first known use of the word. By the end of the decade, Chappe's towers connected many of the major cities in France and by the time Morse transmitted his Biblical exclamation, France had 533 towers covering 5,000 kilometers.[3]

Despite the optical system's advantage over the runner or equestrian, adverse weather conditions often obstructed visibility, impeding its reliability. More important, a new electrical technology emerged that quickly made it obsolete. In 1600 the English physician William Gilbert (1544-1603) had coined the term "electricity," from the Greek word for "amber," the fossil resin used to make jewelry that produces static electricity when rubbed. It took two centuries to move from establishing the term to understanding its scientific principles and harnessing its energy. In 1800 the Italian physicist Alessandro Volta (1745-1827) invented the first battery, the voltaic pile, layers of dissimilar elements that produced a chemical reaction, resulting in a steady flow of electricity. Twenty years later the Danish physicist and chemist Hans Christian Oersted (1777-1851) observed that electrical current induced a magnetic field. In 1825 the British physicist William Sturgeon (1783-1850) constructed the first electromagnet, a piece of iron wrapped with insulated wire that became magnetic when a current of electricity passed through it.[4]

Numerous inventors sought to use the new technology to engineer an electric telegraph, which emerged simultaneously and independently in the United States and Great Britain. Samuel F. B. Morse had established a reputation as a well-regarded painter, but he also dabbled in several mechanical schemes to strike it rich. In 1832 as he sailed home from France following a painting excursion, he learned of the telegraphic experiments from some of his fellow passengers. He became obsessed with the concept and tinkered with the implements, while he continued to paint and teach. In 1835 Morse became a professor of art at New York University, where he received much needed assistance. He did not understand the science of electricity and experimented using only one battery and a small electromagnet. A colleague, chemistry professor Leonard Gale, recommended a battery of twenty cells and an electromagnet with one hundred turns of wire. With the upgraded arrangement, Morse was able to send a current through ten miles of wire coiled around a spool in his classroom. In 1837 he and Gale applied for a U.S. patent. That same year in England, Sir Charles Wheatstone constructed a fourteen-mile electric telegraph between London and Bristol.[5]

In the United States, Morse constructed a machine consisting of a wooden frame that held an electromagnet, a roll of narrow paper (e.g., a cash-register tape), and a pencil strapped to a metal rod that hung down like a pendulum. When he opened and closed the circuit, the electromagnet pulled the pencil down to the moving paper, leaving squiggly marks of various lengths depending upon how long the circuit remained open: a short time produced a "dot" and a longer time a "dash." He rejected the idea of a separate wire for each letter of the alphabet; initially, he thought in terms of having the dots and dashes represent numbers, which would correspond to words, but that system would have required a large dictionary to translate the message.[6]

In 1837 Alfred Vail, a wealthy student of Morse, became his much-needed business partner and helped him improve the machine and the code. Scholars credit Vail with the invention of the "key," the lever used to open and close the circuit, with substituting a pen for the pencil, and with helping to create the code. For his efforts, he received a one-fourth share of the patent; however, his name drifted into obscurity as Morse demanded all the glory and gave no credit to his indispensable associates. Their code matched a catalog of dots and dashes to the alphabet; it copied the arrangement of letters in a box of printer's type, with the most-used letters receiving the shortest assortment of markings (a single dot represented "E"). Quickly, experienced operators developed "sound reading," spelling out messages by listening to the machine's clatter without waiting to read the tape. Subsequently, the pen and paper became outmoded, and the electromagnet attracted an iron bar against a post, thudding "dits" and "dahs" for operators to decipher.[7]

At first, Morse could not create an interest in the new technology. He demonstrated the machine to Congress in 1838, but the lawmakers exhibited little interest; he garnered a similar response from several European governments the following year. Finally, in 1843, Congress authorized an appropriation of $30,000 to construct a line from Washington, D.C., to Baltimore, Maryland, alongside the tracks of the Baltimore and Ohio Railroad (forty miles). Miss Annie Ellsworth, daughter of the U.S. Commissioner of Patents, had delivered the good news of the public funding; Morse asked her to choose the first message. Despite the success of the experiment, Congress did not perceive how it could use the contraption and it allowed private companies to develop it; they rapidly overwhelmed the apathy and by 1850, twenty firms had constructed twelve thousand miles of telegraph lines.[8]

In 1847 during the initial boom, Edward Charles Creighton (1820-74) observed Irish-American contractors erecting a telegraph line near Springfield, Ohio. He was the son of James Creighton, an Irishman who had migrated to the United States in 1805, and who had married Bridget Hughes, an Irish-American, in Philadelphia in 1811. James brought his mother, sister, and five brothers to America and by the middle of the decade, they settled in Ohio. All the siblings became successful landowners. James and Bridget had nine children, six boys and three girls. Edward, the fifth child, attended public elementary school and at about the age of fourteen secured employment as a cart boy in central Ohio in conjunction with the construction of the National Road. When Edward was eighteen, his father gave him a wagon and team of horses, and he began hauling freight between Cumberland, Maryland, and Cincinnati, Ohio. Obviously possessed of enormous entrepreneurial drive, he promptly branched out to supplementary teamster pursuits: road construction, street grading, and railroad roadbed preparation.

Edward's encounter with a telegraph-construction crew of a company owned by Irish immigrant Henry O'Reilly presented him with another opportunity. He began by freighting poles for a line connecting Springfield to Cincinnati and ultimately Louisville, Kentucky. Within months, he became superintendent of construction for a link between St. Louis and Alton, Illinois. During 1848-55 Edward oversaw the erection of routes connecting the major cities of New York, Ohio, Indiana, Illinois, Kentucky, Mississippi, and Louisiana. Through these responsibilities, he became an associate with the powerbrokers that established the Western Union Telegraph Company, Hiram Sibley (1807-88) and Ezra Cornell (1807-74). They combined their eastern companies in 1856 and set about eliminating competition through acquisitions and mergers.

Meanwhile, Creighton maintained his ancillary businesses, and in 1855-56 undertook roadbed construction for a railway right-of-way near Mexico, Missouri, and street-grading contracts in Toledo, Ohio, and Keokuk, Iowa. While his crews attended to those tasks, he performed a "lemon squeezer" for Sibley, the first president of the Western Union. It consisted of a specious survey of a route for a competing line linking Cincinnati to New Orleans. Creighton's activity convinced the owners of the existing wire to avoid the destructive competition of the phantom parallel line and to provide Western Union with favorable rates for its use. During the bogus survey, the newly elected council of Keokuk cancelled the grading contract. Edward directed his youngest brother, John Andrew Creighton (1831-1907), to sell the equipment and to proceed to the newborn town of Omaha, Nebraska Territory.

On June 10, 1856, Edward joined his brothers James and Joseph and his cousins Harry and James "Long Jim" Creighton (all four worked for him) in Omaha. After a short visit, he returned to Dayton, Ohio, to marry Lucretia Wareham on October 7, 1856. The newlyweds journeyed to Pittsburgh, Pennsylvania, to purchase a supply of lumber and returned with it to Omaha on a steamboat, traversing the Ohio, Mississippi, and Missouri rivers. Immediately, Creighton established himself as the city's wealthiest citizen, possessing $25,000 in capital from the sale of the surplus lumber and the Keokuk grading equipment and the income from his other ventures. Edward relocated his headquarters to Omaha in expectation of constructing a telegraph line to

California, which had entered the Union in 1850. However, the Panic of 1857 disrupted the economy and the sectional acrimony that produced the Civil War prevented Congress from passing an appropriation.

While the nation continued to suffer the consequences of the financial panic, the local economy quickly turned sharply upward with the discovery of gold in the Colorado Territory. The Pike's Peak gold rush of 1858 presented extraordinary opportunities for outfitting, freighting, and finance. With his available capital Edward launched new endeavors in all those areas, including a partnership with Augustus and Herman Kountze that expanded their banking operation to Denver and Central City, Colorado. Moreover, the regional prosperity included the stringing of additional telegraph lines. The records are vague, but Edward may have contributed to erection of the lines from Jefferson City, Missouri, to Fort Smith, Arkansas, and from St. Joseph, Missouri, to Omaha.[9]

At long last, on June 16, 1860, Congress acceded to the appropriation of $40,000 per annum for ten years to construct a transcontinental telegraph: "An Act to Facilitate Communication Between the Atlantic and Pacific States by Electric Telegraph," commonly called the Pacific Telegraph Act.[10] During the summer Edward and his associates shuttled between Omaha and Denver, appraising the route and negotiating contracts for poles, while building a line from Omaha to Fort Kearny. Subsequently, he undertook a dangerous unaccompanied investigation of the route to the West Coast, following the trail of the Pony Express. On November 18, 1860, he left Omaha by stagecoach, stopping at Fort Kearny and Fort Laramie and arriving at Salt Lake City in mid-December. He secured the cooperation of Brigham Young, the leader of the Mormon settlement, then rode a mule six hundred miles to Carson City, Nevada, stopping at Pony Express stations along the way, thence to San Francisco by stagecoach. There he met Western Union partner Jeptha Wade, who had succeeded in merging four feuding companies that would build the line east through the Sierra Nevada Mountains.[11] Wade and Edward returned to New York City by sea, which included a horseback ride across the Isthmus of Panama. Edward proceeded to

Dayton to collect his wife, who had returned home to give birth to a son, Charles David, on April 4, 1859 (tragically, the boy died from an undiagnosed ailment on April 12, 1863). Creighton completed his arduous seven-month journey on May 25, 1861, arriving with his family in Omaha aboard the steamboat *West Wind.*[12]

During his absence, on January 11, 1861, the Nebraska Territorial Legislature authorized the incorporation of the Pacific Telegraph Company; most of the incorporators, including Edward Creighton, were Western Union executives. Creighton arranged to have W. B. Hibbard construct the line from Fort Kearny to Julesburg, while he would superintend the operation from Julesburg to Salt Lake City.[13] The work entailed finding the correct-sized trees to trim into poles, no easy task on the treeless plains (teams ranged as far as two hundred miles from camp following stream beds). Workers using long-handled spades dug a hole to the depth of five feet and then raised the pole in the hole, approximately twenty-five per mile. Wire stringers attached un-insulated galvanized iron wire (copper was too expensive) through glass insulators on the poles. A fifty-volt wet cell battery provided the power and could push the current up to five hundred miles because of the minimal leakage due to the low humidity of the Great Plains.[14]

While the Civil War erupted in the East, the first poles went in the ground on July 2, 1861. Making rapid progress, Edward reached the Mormon capital three months later on October 17. The Congressional subsidy included a bonus for speed; the Pacific Telegraph Company earned extra money by arriving at the midpoint first. Edward sent his wife a telegram reading: "This being the first message over the new line since its completion to Salt Lake, allow me to greet you. In a few days two oceans will be united." A week later, the Overland Telegraph Company of California completed the circuit from Carson City; on October 23 Stephen J. Field, the chief justice of California, wired Abraham Lincoln to assure him of Californians' "loyalty to the Union and their determination to stand by its Government on this day of trial."[15]

Creighton became general superintendent of the Pacific Telegraph

Company, which included responding to disruptions in service caused by Indian vandalism, roaming buffalo (who knocked over the poles while using them as back scratchers), and prairie fires. It also entailed building branch lines; from Julesburg to Denver and Central City in 1864; Denver to Salt Lake City in 1866; Salt Lake City to Virginia City, Montana, in 1867; Helena, to Fort Benton, Montana, and Laramie, Wyoming, to Promontory Summit, Utah, in 1869. In the latter year, he also constructed a railroad-specific link for the Union Pacific from North Platte, Nebraska, to Promontory Summit. He concluded the last two assignments after he had stepped down as superintendent in 1867, although he remained on the Pacific Telegraph Company board of directors.

In 1863 with the Kountze brothers, Edward founded the First National Bank of Omaha and became its first president; they also incorporated the Colorado National Bank in Denver and the Rocky Mountain National Bank in Central City. Moreover, Edward remained prominent in freighting, especially long-distance, thirty-plus wagon trains to Denver, Salt Lake City and Virginia City, Montana. He graded a roadbed for the Union Pacific in Wyoming and cut through a bituminous coal vein near Rock Creek; he established a mining company and sold the coal to the Union Pacific. In 1869 he organized the first longhorn cattle drive from Texas to the new UPRR railhead at Ogallala, Nebraska, and became the largest cattleman on the northern Great Plains. He also grazed oxen, horses, and sheep. That same year he helped organize the Omaha and Northwestern Railroad and became its president. Finally, Creighton invested heavily in real estate: three office buildings and thirteen lots (valued at $139,000) in Omaha; thirty-seven pieces of property totaling 900 acres in Douglas County; and 3,757 acres in seven contiguous eastern counties.

Edward Creighton collapsed in his office on November 3, 1874, and died at home two days later at age fifty-four. As he was laid to rest at Holy Sepulchre Cemetery, virtually all business activity in Omaha paused out of respect for its leading citizen. Before his death, a colony in Knox County decided to name their settlement in his honor. In 1958

Edward Creighton, left, is present for a meeting of the directors of the Union Pacific Railroad aboard a private railway car. Seated at the table are: Silas Seymour, consulting engineer, and Sidney Dillon, Thomas Durant, and John Duff, directors. Photograph by Andrew J. Russell. Union Pacific Museum

he was elected a charter member of the Cowboy Hall of Fame in Oklahoma City, Oklahoma. In 1965 Ak-Sar-Ben (an Omaha civic organization) and in 1982, the state of Nebraska, selected him for membership in their respective halls of fame. His wife bestowed his most significant memorial, providing a bequest for the creation of Creighton

University in 1878.[16]

By the time of Edward Creighton's death, the man who had served as his personal assistant during the construction of the transcontinental telegraph had also become one of Omaha's most prominent citizens. Charles H. Brown was born on April 12, 1834, at Stephentown, New York. His father, Randall Adam Brown, a descendant of a *Mayflower* pilgrim, and his wife, Margaret Sweet, had seven children. Their older sons, James Jay Brown and Randall Adam Brown, Jr., migrated to Omaha in 1856 and became teamsters for Edward Creighton; working with John A. Creighton, Edward's youngest brother, they made several long hauls to Denver. Subsequently they began wholesaling to traders dealing with Native Americans and outfitting pioneers moving west. They established a mercantile store at Fourteenth and Douglas streets that became a prominent Omaha enterprise for the remainder of the nineteenth century.[17]

Charles H. Brown followed his brothers to Omaha in 1860. He had graduated from Williams College (1858), studied law at the firm of Seymour and Van Santvoord of Troy, New York, and been admitted to the New York Bar shortly before his departure. He secured employment with Edward Creighton driving ox wagons to Denver, then became his personal assistant in conjunction with the building of the transcontinental telegraph. Following that historic venture, he returned to Omaha and clerked in his brothers' store until October 1862, when he gained election as the prosecuting attorney for Douglas County, the beginning of a prominent political career. In succession the voters made him a member of the Nebraska constitutional convention (1864), a member of the lower house of the territorial legislature for the tenth and eleventh sessions (1865,1866), a member of the Omaha City Council (1865), mayor of the city (1867), member of the state constitutional convention (1875), and member of the state senate (1876, 1878, 1882).[18]

Subsequently, while remaining active in Democratic Party politics, Brown retired from office and practiced law. On June 10, 1886, at the age of fifty-two, in Chicago he married the widow of his younger brother, Mrs. Louis (Eunice Dora) Brown, who had a twelve-year-old daughter, Margaret.[19] A decade later Eunice became a widow again when Charles died on April 26, 1897, fourteen days after his sixty-third birthday, as a result of heart disease. He left instructions for no "ostentation" at his funeral, which took place the day after his death from his home at Twenty-second and Capitol streets with interment at Forest Lawn Cemetery.[20]

Editorial Procedures

Charles Brown had the benefits of a good education and his writing style is sprightly and coherent. He sometimes omitted punctuation, such as apostrophes or periods, which the editors have silently inserted as necessary. Brown's occasional variant spelling of proper names (e.g. Julesburgh for Julesburg and Holliday for stagecoach magnate Ben Holladay) and his inconsistent capitalization have been retained throughout. "Strikethroughs" in the original manuscript have also been retained. A few words added or corrected by the editors and minor explanatory material not found in the notes appear in brackets. Brown mentioned several individuals or localities that could not be further identified.

Telegraph poles are visible behind this 1866 group of emigrants in Echo Canyon, Utah, about fifty miles from Salt Lake City. NSHS RG1227-4-1

Details of *The New Naval and Military Map of the United States* (1862), showing the Omaha-to-Salt Lake City portion of the transcontinental telegraph route. The map incorrectly shows the telegraph route following the North Platte River west from its confluence with the South Platte. In fact, the telegraph line followed the South Platte west to Julesburg, Colorado, and then turned north to follow the North Platte into present-day Wyoming. (The boundaries of Utah, Idaho, and Wyoming have changed since this map was drawn.) Library of Congress Geography and Map Division

Eastern map section, Nebraska and Dakota (Wyoming)

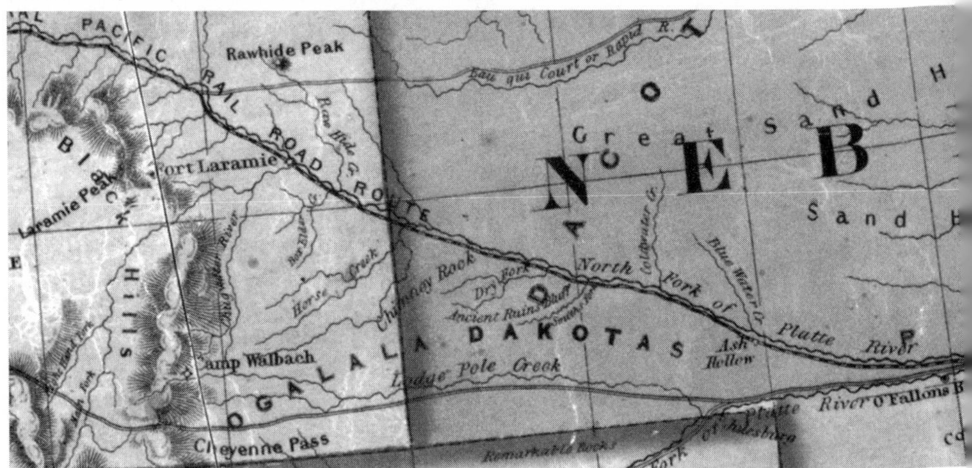

Western map section, Utah and Dakota (Wyoming)

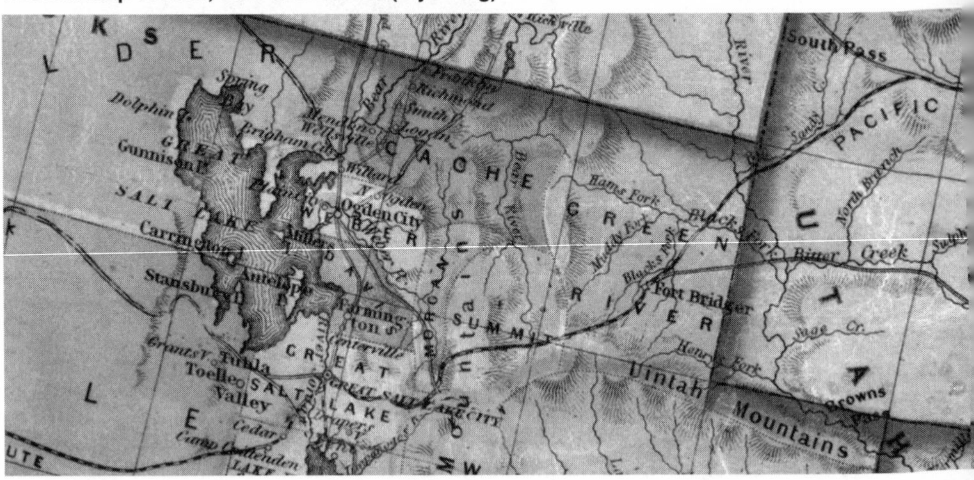

Charles Brown's Diary

My Experiences on the Plains in 1861 in assisting in the Construction of the first telegraph line across the Continent.

In 1860 Omaha had the advantages of being connected via St. Joseph Mo., by telegraphic wire with the great East. This expression embraced that part of our common country east of the Mississippi. During 1860, a line of wire was stretched from Omaha to Fort Kearney two hundred

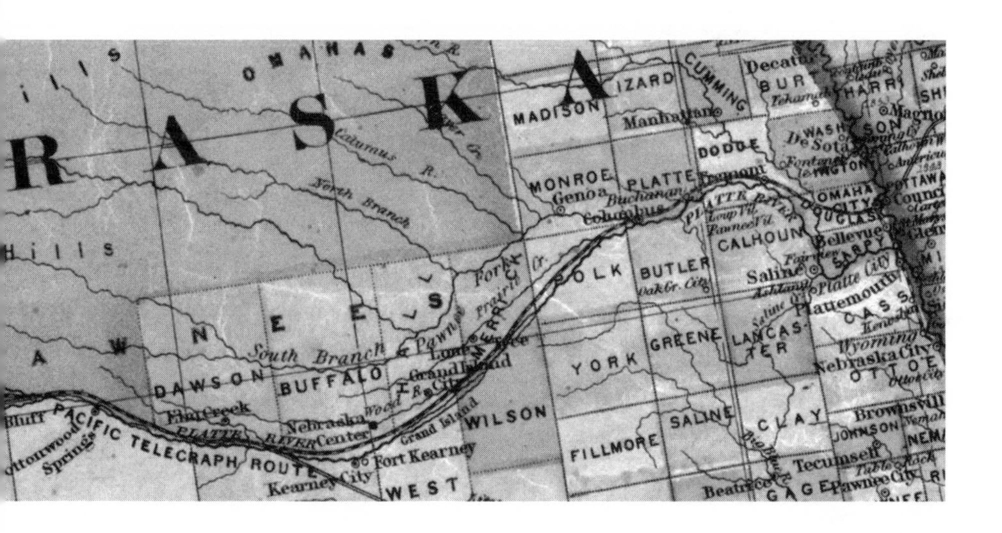

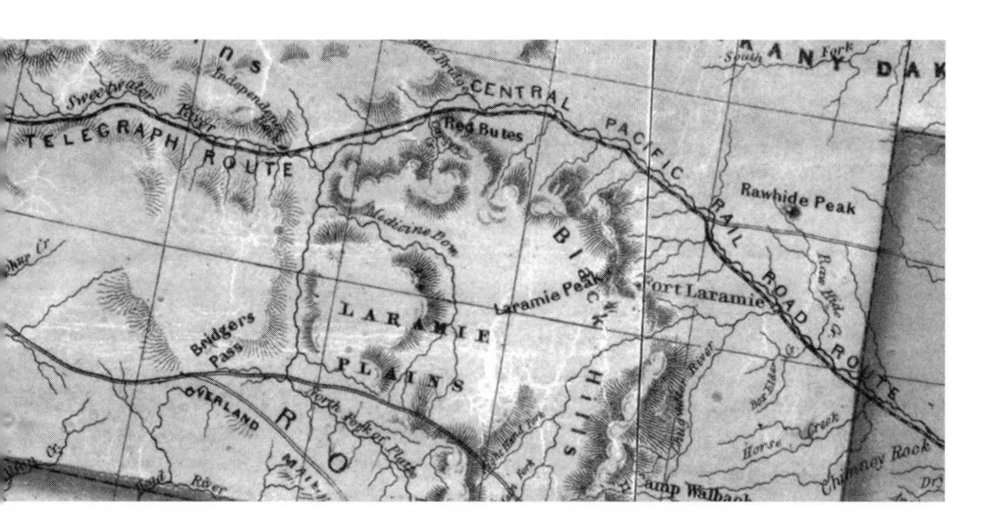

miles to the west. In November of that year Edward Creighton started from Omaha to San Francisco for the purpose of making arrangements with the moneyed men of California for the building of a telegraph line, from Sacramento—then the fartherest point east, in that state, wire had been strung—to Fort Kearney, following the old overland stage route.[21]

Mr. Creighton went by coach on the stage road via South Pass, Salt Lake City, Humboldt River, Carson City (Nevada), and Sacramento. It was a perilous journey to make in dead winter, but he was a resolute, enterprising man, who, having formed his plans, commenced at once their execution.

With not a few hardships he made the trip, a distance of about 2,000 miles. At San Francisco he was successful in the object of his visit. The agreements were that the California party was to build eastward from Sacramento to Salt Lake City while Creighton and party were to build westward from Fort Kearney to Salt Lake City. The length of the line to be constructed by eastern capital was about 1,000 miles; that by the Californians about 775. From Old Fort Kearney to Ogden (50 miles north of Salt Lake City) there are 837 miles; and from Ogden to Sacramento there are 744 miles of U.P.R.R. Co. line of road, as stated in its table of distances. This route, however, is much nearer being a straight line than the overland stage route was, along side of which the telegraph poles were set.

As soon as Mr. Creighton had satisfactorily terminated his business at San Francisco, he at once started for New York City, by the easiest and most expeditious way. He went by steamer from Frisco to Panama and from there crossing the Isthmus to Aspinwall, and from thence to N.Y. City by steamer.[22] He used to relate that in crossing the isthmus he had not a few trials. One of these was his attempt to eat cooked monkey, of which animals the woods, there, then abounded.

After Mr. Creighton had reported to his associates in the enterprise and they had approved of his acts, he set himself about arranging for the funds with which to carry out the work. This he accomplished. Success thus far having attended all his efforts in this direction he returned to

Omaha in March 1861 and made it the point from which to prosecute this great work.

Within a few days after his arrival he and I entered into a contract for my services until the completion of the work of construction. Under this it became my duty to keep his books, pay out money, write letters, and do anything he should require me to perform. My compensation was to be $50 per month. Having nothing to do, I at once entered his employ.

The first thing to be done was to secure the manufactured material to be worked into the line and the supplies to be distributed at points along it, to operate it when completed. These were to be ordered and delivered at Omaha at the earliest day possible. Creighton having the right to send messages over the lines free, used the wire freely. His associates in the undertaking, residing at Cleveland and other cities in the East, with promptness saw that his orders for wire & c were attended to, so that when navigation on the Missouri River opened the first boats up from St. Joseph began to discharge the material on the bank at Omaha. When the heavy steam whistle announced the arrival of a steamer, I had to go down to the place of landing and see if there was any freight aboard for us. If there was I directed where it was to be deposited on the bank of the river and checked up the bills of lading. There were others who sometimes did this or gave me assistance in doing it.[23]

This freight did not embrace many articles. There were the large coils of wire on rudely constructed reels; the insulators, bulky wooden cupped things, long since gone out of use, packed in hogsheads and larger casks; the large carboys, large thick glass bottles, covered with heavy wicker for protection, filled with chemicals for the batteries; ~~and~~ boxes containing instruments and other things necessary for use in a telegraph office; kegs of nails for fastening insulators to the poles; shovels of several kinds; pick axes; iron bars; axes for chopping & c & c.

During all this time, men were selected to take charge of the different trains to be employed in transporting this material. These men were called wagon masters or captains of the trains. Drivers of teams were hired day after day. These men were selected also with reference to their ability to swing an ax. Laborers for digging holes and setting poles were secured. Cattle and wagons were being purchased. A large herd of cattle and twenty-eight wagons from Lillian Springs were sent for and brought to Omaha to be used on this expedition. They came in, in good season. Extra care was taken to have every wagon in the best state of repair possible. Camps were formed at which the cattle and wagons were taken and the men gathered. All were kept steadily at work.[24]

As the wagons were repaired, inspected, and pronounced sufficiently strong, they were taken to the river bank and loaded. Each one was numbered and a list of the articles put on board kept and also their weight. The number of pounds in each wagon differed. Some had aboard as high as 7,800 and the lightest laden one 4,200. The bulk of weight was in the coils of wire, which was heavy and galvanized to prevent its corroding. The insulators were taken from the casks in which they were shipped and packed in the wagon boxes as compactly as possible. To this extent the freight was lessened. Considerable time was consumed in loading.

We loaded by trains and as each train had its wagons filled and was properly provisioned and otherwise well outfitted it rolled out and took up its march over the plains and into the mountain region. Sometimes our work of loading was delayed by reason of having to wait for freight to come up the river on the steamers.

The following were the trains engaged in hauling the freight:[25]

James Creighton's with 30 wagons, two of which were loaded with provisions and camp equipage & c. His wagons were drawn by from four to six yokes of cattle to a wagon.

George Guy's with 3 wagons, two span of mules to each.

John and David Hazard's with 15 wagons with 3 or 4 yokes of oxen to each.

Joseph Creighton's with 10 wagons with 2 to 4 yokes of oxen to each.

James Dimmock's with 5 wagons each drawn by three yokes of oxen.

Aaron Hoel's with 20 wagons each drawn by 3 to 5 yokes of oxen.

Matthew J. Ragan's with 5 wagons each drawn by from 2 to 3 span of mules.

Robert Tate came from Denver, Colorado, with three wagons each drawn by three yokes of oxen and joined and was merged with Joe Creighton at or near Scotts Bluff.

Ragan was the first man to move his train. He had a contract to deliver as many poles as he could at Julesburgh, the point at which Creighton commenced his work of construction. This place is distant from Omaha about four hundred miles. Creighton had arranged with W. B. Hibbard to build the line from Ft. Kearney to Julesburgh, while he took charge of and had general supervision of that portion of the line to be built west of this point. Hence what I write will in no way relate to the construction of the 200 miles of line between the last two mentioned places. The poles, which Ragan was to deliver, were to be used in the line west from Julesburgh. He cut them in the bluffs south of Cottonwood Springs. They were cedar and of course durable, but small and far from straight and shapely. They were the best that could be had and accept them we had to. Ragan delivered poles enough to construct the line from Julesburgh to a point about ten miles west of Old Fort Laramie, excepting about 20 miles west from Mud Springs, which Dimmock and myself with his men cut and distributed. Ragan had to haul his poles one hundred miles before delivering them, and he returned to the cañons with unladen teams.[26]

The next train to move out was James Creighton's. With the loading of his wagons I had little to do save keeping record of the materials stored away in each wagon and their weight. He is a man who does his own business. I think he started about the 10th day of May. He encountered no difficulties until he reached the Raw-hide bottoms. Here the passage of the wagons became almost impossible. This small stream had overflowed its banks and the ground was most thoroughly saturated with water and consequently soft and incapable of sustaining the weight of the wagons. The wheels cut through the sod and sank to the axles. Sometimes in doubling up the oxen there were as many as 20 yokes to a wagon. This process of movement was exceedingly slow and he could only make about 2 miles per day. It is needless to say that at times the oxen while pulling on the wagons became mired. To lay still was loss of time and involved expense. He resorted to the expedient of covering the ground with prairie hay. Load after load of this he purchased and used. It served well the purpose. The road was paved with hay. Of course at times the wagons would break through it and took much man and cattle labor to extricate them. After the train had passed over the hay, it was gathered up and spread on the ground in advance. It took considerable time to get over the Raw hide bottoms. From there onward up the Platte Valley the roads were wet and heavy and the progress of the trains slow until Silver Creek was reached, where by reason of the overflow of this creek and the Platte River the grounds were water soaked and the same difficulties had to be overcome and in the same manner as at Raw hide.[27]

At Plum Creek about 34 miles west of Ft. Kearny one of Creighton's men was accidentally shot and, after a painful lingering for several hours, died. They were in camp, and this man and another were going out hunting. They went to their wagons to get their guns. The man who was shot stood on the wagon tung [*sic*]. The other was in the wagon and heedlessly taking hold of the gun barrel pulled it towards him. The gun hammer caught upon something was raised and when it fell it exploded the cap and the contents of the gun passed into the man. Death at any time brings consternation and sorrow, but when it occurs in a camp of men, with no physician to minister to the sufferer, removed from all contact or means of communication with the being world and under the circumstances of this man's taking off, it is simply horrible. A grave was dug and using a wagon cover as a winding sheet shroud they laid him in his grave, to take that long sleep which knows no waking, till all are summoned to a final judgment. With sad hearts, the train moved from this camp of death.[28]

The third train to start was Hazard's. To the loading of this I gave considerable attention. Creighton (Edward) had hired, at low wages, some Mormons, mostly young Englishmen, who were on their way to Salt

Lake, to work for him. Their compensation was small yet they were in fact the dearest laborers we had. They were from the cities and factory towns of England and had with one or two exceptions, no adaptability to the work they were to do. They could not learn to be handy drivers of oxen, nor to do well and with ease the kind of work for which they were employed. These young men were assigned to Hazard's train.

An amusing incident transpired at the time the cattle were being yoked for the purpose of starting the train on its journey. One of these English boys was given a yoke and told to yoke up two oxen pointed out to him, patient, gentle animals. He shouldered the yoke and went up to the off-ox and tried to slip the bow, not taken from the yoke, over his head and horns. This attempted feat created great merriment [Pages 11-15 are missing from the original handwritten manuscript.][29]

[June 16, 1861]

[Creigh] ton's train. The driver stopped and held the stage a short time for me to examine the train and ascertain its needs, if any. Nothing was wanted. Joe had made very slow progress, about 10 miles per day. I insisted he should make from 15 to 18 miles each day.

With Joe I found my cousin Henry J. Sweet who, with his partner, was taking two wagonloads of freight out to Denver. Henry was driving two yoke of oxen to his wagon. He was dirty and looked very much like a bull-whacker. Bidding him goodbye I jumped into the coach and away we went at the crack of the whip.

At Fremont, about 40 miles on our way, we made a shift of horses. As soon as the stage came to a stop my fellow passengers and myself concluded to walk forward until overtaken. The walk rested us very much.

There were four of us in the stage, Kimball, Wilson, Davenport and myself, and we soon became if not acquainted, at least quite sociable. Kimball was from Winterset, Iowa, and on his way, as were the others, to Denver. He was a heavy set man, abut 40 years of age with good common sense and companionable. Wilson was cast in a different mould. He was about 33 years of age, medium height, thin and spare, sharp black

Charles H. Brown, from J. Sterling Morton and Albert Watkins, *Illustrated History of Nebraska,* Vol. I, third edition (1911).

eyes, heavy black whiskers and partially bald. He had been a book keeper for the last 8 years in a branch bank of the Iowa State Bank at Oskaloosa, Iowa. He was well informed, but not over talkative. He was entertaining and when he spoke said something. Davenport was quite the reverse of these two men. About 27 years of age, a little above average height, well proportioned, black hair and eyes, with fancy gauntlets on his hands, wearing checked pants, a stunning neck tie and Byron collar he had the appearance of what was then called a swell, now a dude. He was inclined to be garrulous and boastful, and evidently prided himself on his own importance. He claimed to be a married man, and exhibited a likeness of a handsome young lady whom he called his "Yankee girl."

We spent the day in conversation and in reading a novel. When tired of our own talk we turned to the novel and in turn read aloud each a chapter. The name of this work of fiction was: "The Comical, quizzical, tragical adventures of a Hoosier in the Mound City of the far West."[30]

This day was excessively hot, although the strong current of air passing through the coach prevented our suffering very much. At night we were terribly annoyed with the mosquitoes. Having slept none the night before, I was very much fatigued with the day's ride. However I slept very little during the night as the mosquitoes and the jolting of the stage prevented. Night wore itself away slowly and the morning of

June 17th [1861]

came upon four weary, dirty travelers. We stopped for a change of horses and for breakfast at "Pap Lamb's" famous ranch. After a good wash we partook of one of Mother Lamb's best meals. In those days she was celebrated and most justly too, for her good cooking.[31]

At this place another passenger joined us. She was of the female persuasion. She was neither fat nor forty, but fair and about twenty-three. She was neither bashful nor diffident and had a wagging tongue. We all talked to, at, and with her and by degrees drew out her past history. Her parents and sister were Mormons, but she was <u>not</u>. Of course not! This she was desirous we should distinctly understand. At a social game of cards she was my partner and she played an excellent game. As gentlemen we were bound to believe she was no Mormon, but notwithstanding this she left us at Joe Johnson's ranch!! Joe Johnson is a Mormon and is notorious in the possession of five wives by the grace of Brigham Young. He has been a resident of Salt Lake City but is now living at Wood River and edits and is the owner of a paper named: "The Huntsman's Echo." This paper is ably conducted for a country journal. Joe is a smart, shrewd active man but, if reports are true, somewhat unprincipled and unscrupulous in his dealings with his fellow men. He writes with vigor and in a sledge-hammer style. He is a character and his name will live in tradition long in this part of Nebraska.[32]

We arrived at Kearney at 2:30 p.m., thus making the trip of 200 miles in about 36 hours. I was very much fatigued. The stage station is north of a line drawn from Ft. Kearney to Kearney City or Adobe Town and about equidistant from these places. At the station I wrote two letters.[33]

In September 1862 I was again at this place. I went up to serve upon Ben Holliday a legal paper styled "ne exeat regno" and was specially deputized for this purpose. I have forgotten who was the plaintiff in the suit but J. M. Woolworth, Esq. in whose office in the winter of 1860-61 I read law was his attorney.[34] Holliday was on his way to the States from an inspection of his overland stage routes. I hovered around the station for two days waiting the arrival of Mr. H. During the afternoon of the last day the coach in which he and some of his attendant friends rode came up to the station with the four horses at full running speed, leaving a cloud of dust behind. Four men jumped out and went into the dining room. After inquiry I learned which one was Holliday. I waited at the front door, from which I supposed the coach would depart, for him to appear when I was to hand him the writ. To my astonishment in a short time I heard the loud crack of the driver's whip and quickly following, the rumble of the coach wheels and then instantly from behind the station there came the overland coach and on the road eastward it went as fast as the horses could jump. I had lost my prey. Ben had come and gone and I had seen him that was all. I was dumfounded and felt awfully chagrinned over my failure to secure service of the writ. The team was hitched up at the stables back of the station and Ben and his company walked out of the dining room and entered the coach and departed leaving me watching for him on the other side of the building. There was no intention on his part to elude me but he did all the same. After my inglorious failure I then went back by stage to Cottonwood to find a place to cut hay on a government contract which my brothers and a Mr. Black had.

While I am thinking of the celerity in which Holliday came and went in "a cloud of dust" I am reminded of what Mark Twain, in his "Innocents Abroad," makes one of his travellers in the East say to his guide who was telling him about the desert, in which the children of Israel wandered and said, "It took them forty years to cross it." "What?" said the American traveller, "Forty years? Why Ben Holliday in his overland coaches, would have taken them across in forty hours!"

Fort Kearney was built, if I remember correctly, in the year 1847.[35] It has about a six company capacity. Since its establishment, soldiers have been constantly quartered in it and have served to hold in check the Indians and to a great extent prevented them from committing depredations upon the overland emigrants and later the settlers of the lands to the east of it. They have had more of a moral than a physical restraint upon the aborigines. A "Fort Kearney Man" as the Indian calls a soldier, is regarded by him as a dangerous man and ~~the mention of one~~ the sight of one causes more fear to an Indian than does the sight of a good stout warlike saint to the devil.

The Indians used to call me a "Fort Kearny Man" as the sight of my left hand led them to believe I had been wounded while serving as a soldier.[36]

At the time which I am writing about there was only one company at the fort, and that of dragoons. All others had been ordered to Fort Leavenworth to be in readiness to march quickly to the aid in the suppression of the rebellion.[37]

The Western Stage Company's line terminates at this place, but Holliday runs a daily coach to Denver and a weekly line to California via Ft. Laramie, South Pass, ~~and~~ Salt Lake, and Virginia City in Nevada. The company is doing a large and lucrative business.[38]

About 5 o'clock I walked up to Adobe City, a distance of three miles. Here I found in camp two men from Omaha on their way to Denver. Charles E. Parcell and Doc Smith, not our Doc Smith, the surveyor and poet, who sends forth his philosophical poetical effusions from "Dox Box."[39]

The inhabitants of this little mud house city would not be worthy members of a church society. They are preeminently hard cases and "tough cusses." In nearly every building in the town there is a saloon with a small assortment of groceries as a side show. The chief occupation of the men is gambling, horse and cattle stealing, and drinking whisky, while they advertise for respectability by large signs reading with variations like this:

Wholesale and Retail Dry Goods and Groceries
Great Outfitting Emporium
Last Chance to Get Supplies
Goods sold for Cost
Wines Liquors of all kinds and Tobaccos
Cheap Emigrant Store Saloon
Pay John Talbot[40]

A few days previous to my arrival, Ragan who was to get out poles at Cottonwood, had eleven head of very fine mules stolen from him while he was in camp a few miles to the west. Up to the time of my arrival no clue had been obtained of them.

The country from Omaha to Elk Horn River is rolling prairie. The soil is deep, rich and well adapted to agricultural pursuits. The crops looked splendid. The Platte Valley to this point on the north side of the river is broad and beautiful. Its soil is rich and capable of being made to produce largely of every kind of crop susceptible of cultivation in this latitude. It is being settled upon by the pre-emptor quite rapidly.[41]

This part of the western world receives its greatest drawback from the scarcity of growing timber. There are no woodlands in eastern Nebraska and none at all unless they are to be found in the extreme northwestern part of the territory. As yet no coal mines have been found with a depth of vein to justify working. I have no doubt coal mines exist and will be discovered and worked although geologists in their charts omit Nebraska from the lands having coal deposits.

Most of the rivers and streams are bordered with trees. The cottonwood predominates largely and some of them are very old and of immense proportions. The other varieties are box elder, elm, black walnut, red cedar, soft maple, ash, hickory, basswood, and two or three varieties of oak. The absence of timber in a country like this is not so great an obstacle or prevention of settlement as forest covered lands would be. A settler in a forest has an herculean task before him even to get sufficient soil to cultivate to sustain life for a few years. He must cut and hack his way through the primeval woods. At the close of a lifetime of hardest toil he has only a patch around his dwelling for production. It takes generations to transform a forest to fields of pasture and meadow and luxuriant grain. Not so in this part of the West. The soil is ready for the plow. Even the first year of plowing is made profitable by the planting of corn, which when harvested is called the "sod corn crop." While the farmer does not make any calculation on this crop, there have been fields which have yielded as high as 40 bushels to the acre. As to pasture and meadow land, nature has provided them. The second year's plowing brings the crops. As for fuel it is astonishing how little is absolutely needed. It must be admitted that economizing on this necessary article of consumption is very inconvenient and entails some suffering. The civili[zation] of today rests on the sufferings and death of those who labored to produce.

It is easier to grow timber and fuel in a rich treeless soil than "to clear" the land in a wooded country for cultivation. In a few years a farmer by planting cottonwood sprouts, in a suitable place, can have wood enough for use. This tree is a very rapid grower. The soft maple in growth almost rivals the cottonwood. The farmers have learned this and on their farms are to be found groves of the several kinds of trees. As the country settles, the prairie fires become less frequent and less frequent until they are known no more. When these fires are not, then it is wonderful how rapidly the tree belt along the streams and the groves in the ravines spread. There is more woodland today in Nebraska (1891) and would yield more cords of wood than in 1854, the date of the earliest settlement, notwithstanding the 37 years of consumption and waste.

June 18, 1861

I staid last night at John Young's and slept on the floor in his kitchen. It was an earth floor and I had only a buffalo robe under me, and my own blanket as a covering. When I went to bed I was well worn out from the loss of two nights' sleep and 36 hours ride in the coach. I did some tall sleeping.[42]

The first thing I gave my attention to this morning was to secure a way of going forward to Julesburgh.[43] In looking around I found there was an old man in the place with two teams who was on his way to Denver. He had camped in town last night and was to start at noon. In looking around I found him in Talbot's Grocery-Saloon playing billiards. When I first saw him he was well disguised with liquor. I arranged to go with him. I was to pay him $5.00 for carrying my blanket and tin case. He wanted $15.00. I was to walk and camp with him at night. At noon he was not ready to start, although his teamsters were anxious to move on and importuned him to go on. He was engaged in playing billiards for money. He was betting with the man playing against him. There were ten or a dozen of the "city" loafers and gamblers in the room and they were among themselves doing a large amount of sham betting on the players. This was to blind the old man, who was constantly becoming more drunk. He was loosing [*sic*] although occasionally he was allowed to win a game by his opponent who was a much better player than he.

We were not permitted to leave until 5 o'clock p.m. The old man, whose name was Lesher, kept one of his men with him. His name was Frank and Lesher had great confidence in him. I was inclined to think he was "a sport" and played the old man. How long the game would last was to be determined by the amount of money hazarded at such games. That the old man was to loose [*sic*] was certain.

Considering the lateness of the hour at which we started on our journey we made a long drive and camped for the night at the "Seventeen Mile ~~Ranch~~ Point." Took my supper at Sydenham's Ranch, a splendid meal—bread and milk.[44]

In front of Keeler's Ranch I saw three golden eagles made fast by cords tied around their legs. They were young and I should judge just able to fly. This man Keeler is a brother to the notorious Tom Keeler of Douglas County, who about fifteen years later was shot and killed by Dan S. Parmelee at Elk Horn Station in said county. The keeper of this ranch sustained a very bad reputation.[45]

The day was excessively warm.

June 19, 1861

Went to bed last night at about 10 o'clock and for the first time this year tried to sleep out of doors. Made up my bed on the ground under a wagon. I found it impossible to sleep. My time and energies were given to fighting away mosquitoes. The air was dense with them. Sleep was a failure. I fled from them and took up quarters in Sydenham's ranch, who gave me permission to sleep on the ground floor in one of his rooms. In a short time I found I had made a great mistake in changing sleeping apartments. I not only had the night birds of musical voice and long sharp poisonous bill, but worse and worse, bed bugs in infinite numbers. This I could not endure and again returned and made my bed under the same wagon and I felt content, in a measure, with my birds. About midnight a good strong breeze came up, before which the winged pests disappeared and I had a good sleep and rest.

We started early and at noon camped at Plum Creek, which is distant from Fort Kearney thirty-six miles. It was at this place in the latter part of last November I witnessed a bloody affray between a few Adobe Town boys and the Sioux Indians. The Indians outnumbered the whites at least ten to one. The Sioux were taken unaware and fired upon by the whites, who killed one Indian and his mule to my positive knowledge. These white desperadoes claimed they had killed two other Sioux in the bluffs, into which the Indians precipitatedly fled, followed by their bloodthirsty assailants. In the bluffs the Indians collected and charged upon the whites, who in turn fled in greatest haste before their late retreating foe, and had it not been for the appearance of the U.S. government troops in the field, the Indians might and probably would have exterminated not only these desperadoes, but very many pilgrims also would have suffered death at the hands of the savages. This attack upon the Indians was unprovoked and unjustifiable in every respect. It is a great wonder that the whites are accorded such lenient treatment as they have from these wild men of the plain[s].[46]

An account of this fight I have recorded in my experiences on the plains in 1860. At night we camped at "Irish Tom's."[47]

June 20th [1861]

Slept last night in Tom's ranch on a buffalo robe spread on the ground. During the latter part of the night the lightning was very sharp and thunder heavy but the rain storm passed around us. The lurid flashes of lightning and the deep heavy rolling thunder caused our horses to take fright and pulling their lariat pins started east on a run. Of course there was nothing for the boys, who had in charge the old cub's teams, to do but to give immediate chase. They found them ten miles away. They had been stopped by some campers who tied them up. This was certainly a great favor, but for rendering it and to make it appreciated they extracted $2.50 for their services.

We rolled out of camp at a late hour, owing to the stampede of our horses. The storm last night gave us a cool delightful day and we stopped at Smith's new ranch for nooning.[48] There was an encampment of Indians at this place. There were about fifty of them and they had for shelter five tepes or wigwams. These tepes are made of poles covered with tanned buffalo hides. These poles are from twelve to fourteen feet in length and about three or four inches in thickness at one end and one or two inches at the other. The large ends of the poles are put in the ground in a circle about six or eight feet in diameter and then brought together near the small end and fastened. The skins are drawn around the poles and fastened to them and also are pegged to the ground. A small opening is left in the top of the tepe for the smoke and impure air to escape. The fire when used is built in the center of the conical tent. These portable houses are not warm nor otherwise comfortable. The Indians sleep in them with feet to the center and head to the tent covering. They are not as particular in the occupancy of their houses as the sect of Shakers, for both male and female occupy the same tepe. It is seldom the bucks go on a journey without their families. The buck or male Indian is a professional gentleman. He is a warrior and hunter. He does no work. The squaws do all the work. They are the drudges and slaves of the braves. The buck is too proud or rather too lazy to walk. He rides. When they go into quarters— pitch their tents, the women do all the work. They unpack the ponies, and picket them out. They erect their wigwams, do all the cooking while their lords squat around and do the talking and eating. When camp is broken the females are the ones who do the work. They take down the tents, roll up the skins, pack them and their other portable property on their ponies. Their tent poles are fastened to the saddle on both sides of the pony and the large end of the pole rests and is dragged on the ground extending behind the pony several feet. On these poles and behind the horse I have seen not only large bundles of skins fastened and thus transported, but aged, infirm, or sick savages. They make low wicker-work baskets, from willows and strips of raw or tanned hide, and these are fastened to the poles extending behind the ponies and ~~these~~ are used for carrying the Indians unable to travel and whatever else they wish to place in them. The horses, in addition to drawing the loaded poles, have also to carry on their backs large loads upon which sometimes the squaws take seats and ride. Generally the squaws lead their ponies. They also utilize their dogs by fastening to them small poles on which are bound articles for transportation. The loads they drag seem incredible. These dogs are ferocious looking brutes. They are undoubtedly a cross of dog and wolf, resembling both. I am told this is a fact in corroboration of which I can state I have seen in Indian encampments tame wolves.

The pony is really the Indian's only beast of burden. He is a small, compactly built animal, hardy and possessing great powers of endurance. Their only food is prairie grass. In the spring, summer, and fall months they thrive well on the grasses, but suffer much during the winter months and are very poor when the new grass springs up. The Indian is improvident and makes no provision for feeding his ponies, which subsist the year round by grazing.

These Indians were the Ogalalla Sioux—a band of the great Sioux tribe. The Sioux is the most powerful tribe of Indians in our country. They fraternize with the Cheyennes who own and inhabit this region of country. These two tribes of natives are very splendid specimens of the human race physically and from general appearance mentally. The men are tall, raw boned, athletic. They will average fully six feet in height, while the women are small and usually well built. A small short man or a tall large woman is seldom seen. I do not think I ever saw one among them. I never saw enough of these tribes to form any opinion of their domestic relations.

The Sioux bands of Indians used to inhabit the country stretching from the Rocky Mountains eastward and through the Dakotas, Minnesota, and Wisconsin to Lake Michigan. They spoke one language, though there were tribal dialects. Their warriors numbered thousands and were fierce, brave fighters. Their chief, if not only, occupations are hunting and war. Their pitched battles, in which there is a test of strength of the contending tribes, are very few. Their wars are successions of predatory incursions in which a few scalps are taken and ponies stolen in large or small numbers as success waits on their thievish

"Ogalillah Sioux Village, North Fork of the Platte, Na" (*sic*). Photograph by Albert Bierstadt, 1859. The Oglala are one band of the Lakota, also known as the Western Sioux. NSHS RG3122-1

attempts. The most inveterate enemies of these [the Sioux], the Arrapahoes, and Cheyenne are the Pawnee, who used to occupy Nebraska east of Wolf or Loup river and whose present reservation is north of Columbus and about ninety miles west of Omaha. The Pawnees were not a numerous tribe, but they were accounted great and successful warriors and were capable of contending with the braves of all other tribes.[49]

The Sioux men are beardless. They wear blankets and buffalo robes, as blankets, and tanned buffalo, deer, elk, or dressed antelope skins for leggings which are brought up and girded around their waists. Their feet gear are moccasins. The bucks wear their hair long, done in a large braid and extends from the top of the head backward and down the back. This plait of hair has fastened to it from the pole [poll] of the head to its end large tin or sometimes silver discs, thin and oval, from two or four inches in diameter. The largest one is put on near the top of the head and they diminish in size as they extend downward so that the smallest one is at the end of the braid. The chiefs and other men of distinction among them wear the silver disks. These are pounded out of silver coin.[50]

The squaws wear dresses—very feeble imitations of those worn by their white sisters. They also wear moccasins which they manufacture.

The upper leather is usually tanned deer skin, while the soles are heavy buffalo hides tanned only on one side. These women did not wear their hair long and it was not done up in braids. They have the reputation of being virtuous, but I noticed once in a while that there was a half white papoose among them, which the mother persisted in saying she had found out on the prairie!

This squad of Indians was moving toward Cottonwood Springs, where they were collecting for the purpose of holding a war dance at which was to [be] exhibited the scalp of a Pawnee, taken in a late engagement. Taken all in all the day was a pleasant one. We stopped for the night at Gillman's ranch.[51]

June 21, 1861

Slept last night in the ranch. Was up bright and early this morning and in a short time started on our westward journey.

Quite a number of teams and men were camped at Gillman's and they had a large herd of young cattle, which they were driving to Oregon. They were from Iowa.

Stopped at Cotton-wood Springs post office to ascertain whether I had any mail there. I found none.[52]

Stopped at noon at Dodge's Ranch. Mr. Dodge is a gentleman about sixty years of age and is the father of Genl. G. M. Dodge. He formerly resided near Bridgeport in Douglas County on a farm he owned. He is somewhat eccentric, but nevertheless keeps a respectable "shebang."[53]

At this place were camped Indians and in the afternoon on the way up as far as Jack Morrow's Ranch we were meeting Indians on their way to Cotton Wood Springs to attend their "pow-wow."

At Morrow's met Matthew J. Ragan who with his small number of men was cutting poles for the Pacific Telegraph Company. He had cut and hauled down to Morrow's about five hundred red cedar poles from what are known as cotton-wood cañons. He and his men were in camp in these cañons. In a few days he was to haul them to Julesburgh.

Jack Morrow is a character. He is a well educated man and knows how to act the gentleman. He is an encrusted diamond. He lives in his ranch comparatively in ease and luxury. He keeps a store in his adobe building and carries a stock of merchandise consisting of a few articles in the dry goods line, an assortment of coarse ready made clothing, a good supply of canned goods, groceries of most kinds, especially flour, bacon and molasses, and liquors of all kinds. These "stores" pay well. The ranch men procure their supplies in the main from Omaha and St. Joseph and haul them in wagons to their places of business. Jack at most times is a cold water man, but he has his "periodical drunks." These last him from one to two weeks and during these spells he is a "holy terror." Invariably when in Omaha on his trading expeditions, after he has completed his purchases, he winds up his stay with a spree. I saw him at one time, at a circus, drive up to a barrel from which a man was vending lemonade and take a large sponge from his wagon and dip it into the mixture called "lemonade" and wash his buggy with it. All were amazed at this and no one more so than the owner, but no one said anything or in any way interfered. Having finished washing his wagon he turned and asked, "What's to pay?" He was given a good heavy price for the lemonade. He said nothing, paid the bill, with $5.00 extra, and then turned and drove away. On one or two occasions he has broken up the furniture in saloons, but he always "paid for his fun." He was an inveterate poker player and skillfully played the game and often times won as well as lost "large stakes." In business matters he was honest and had as a consequence, excellent credit. His race was not long and he died in Omaha in fact a poor man.[54]

We went into camp on the open prairie about six miles west of Morrow's. Old Lesher, with Frank whom he kept with him, overtook us here. Lesher was drunk, cross, ugly, and poured out volley after volley of oaths at Tom the driver for not waiting for him. He endured what the old man said with the greatest indifference, all perhaps for the best. He admitted he had lost considerable money at "Dobe Town." I judge they took from him nearly all he had.

He had with his outfit three choice horses, finely bred. One of them

O'Fallon's Bluff, 1866. Photograph by Charles R. Savage. NSHS RG2154-8-27

was a fast trotter as they claimed. They were indeed beauties and the old man adored them.

June 22, 1861

Slept last night in the open air under the wagon. Lesher concluded to remain in camp for a few days and freshen up his animals. Watching for chances to get passage with some other pilgrims, I soon found a man with whom I made satisfactory arrangements. Putting my traps in his wagon I went forward with him. The day was hot but during the forenoon we made the journey over O'Fallon's Bluffs and took our nooning at Bob Williams' hotel, which is at the foot of these bluffs on the west.[55] On the 28th of last October we were driven into camp on these high bluffs by a heavy blinding snowstorm. We, however, found good grounds for our cattle under the bluffs among the willows skirting the Platte River. As it turned out it was fortunate for us that the storm drove us into camp at the place it did, for all the other trains which made over the bluffs had no grazing grounds for their cattle, while our own did splendidly by eating rushes and willows. Stopped for the night at Dorsies [Dorsey's] Ranch. Here I found Cyrus Morton in corral. He was in charge of his own and Harrison Johnson's teams and was on his way to Denver with freight. There were others here—six or eight teams more—and they all made up a nice party, and had a good visit with them. Among them was a young lady from Rock Island, Ills., who was on her way to Denver. She was well educated, of refinement, and fairly good looking. She was on her way to join her husband in Denver. Played cards by moonlight till late at night and after taking "a snifter" of Cy's good old Rye Whisky I bade all good night and retired to rest in my bed in the open air. Had a splendid sleep last night as we were camped on high, dry ground and a good gentle breeze was stirring.[56]

At this camp I saw a rat come out of the front end of one of Morton's wagons and clamber down on the wagon tongue. He jumped on the ground and made for a pail of water, which stood near the fore wheel and proceeded to slake his thirst. After drinking his fill, he started to return to his home in the wagon. One of the men with his pistol fired at him with only the effect to hasten his return to quarters in the wagon. This rodent had undoubtedly stolen his ride thus far, from Omaha, on his way to Denver.

Rats and mice were in 1860 very plentiful and destructive when I was there [Denver]. I was informed they had been brought in wagons from the Missouri River. I was incredulous on this stated fact, but what I saw at Morton's camp convinced me of the truth of the tale told me.

In 1860 John Plumbach, who was a farmer residing about six miles southwesterly from Omaha, took a load of cats to Denver and made a pile of money from their sale, realizing on the best looking ones as high a sum as fifty and sixty dollars.[57]

June 23, 1861

Took a late start. The day was hot and not a breath of air was in motion. Only two events or objects of note worth recording fell under my observation.

One of these took place at a camp which we reached after we had traveled about one hour. A white woman there gave birth to a child and that too in the morning before breakfast. Mother and child doing well.

In the afternoon witnessed one of those rare optical displays called mirages, produced by the heat. It lay off to the west and had the appearance of a small lake with its waves rolling and chasing each other to shore, bordered with trees waving ~~in the~~ back and forth as swayed by gentle winds. It was really a beautiful sight and when seen for the first time is truly deceptive.

Sunday! Today Sunday? Can it be and I here on the plains over three hundred miles from Omaha? Yes, Sunday is one of the seven days of the week and comes and goes with the same unending regularity that marks all the other days. Here to the dusty dirty pilgrim there comes no one day set apart for rest. Nothing tells of its approach, it is truly here only a day in memory and all days remain the same. There is no change. No church bell rings the hour for Christian worship. Above is the broad blue sky with its blazing, scorching sun; under our feet the dry, hard earth. So has it been for years. The ranches are open and trade goes on as usual. Men continue their regular toil, not even ceasing in their swearing and drinking. The trains move lumberingly along as on other days. Sunday on the plains? No! Only in name. Saving this, it has disappeared. Here life is rough and wild; and how easily we glide into savage customs.

Camped for the night at Giroux and Dixon's—the Lone Tree Ranch. These persons married half breed Indians, daughters of the old French trader Beauvais . These women were fair, comely squaws and in matters of dress followed neither the whites nor Indians, but wore clothes about half way between them—a miscegenated dress.[58]

June 24, 1861

This morning had all the glories of a sunrise on the prairies, in a clear cloudless sky. The day commenced very hot and we took a hurried and early start. Barnes, for such is the name of the man I have been traveling with for the last two days, left me at Scott's ranch and returned to his home in Plattsmouth, Nebraska. He was peddling eggs, butter, cheese, potatoes, & c to pilgrims and ranchmen. He had so nearly sold out he was not justified in going further as he could in all probability dispose of what he had on his way back.

This ranch was kept last year by Goddard and Scott. It was here that Creighton (John A.) for whom I was a bull driver for my board last year, discharged one of his men, "Mexican Bill," who had become so ugly that we could not get along with him. He was cross and quarrelsome and we did not know how soon he would prove a dangerous man on our hands, so we turned him off—although he was a most skillful and expert "bull whacker." His equal in this line I have never met. The night after Bill's discharge at this ranch he had a fight with a French man, Leopoldi by name, who was keeping a ranch three miles above. They exchanged a few pistol shots and then drew for close fighting, their bowie knives, but were finally separated before cutting and slashing each other. As it was neither of them received any injury. The quarrel originated over a game of cards at which money was wagered. Bill was adroit in handling knives, especially in throwing them. I have frequently seen him throw them with remarkable precision at objects. He used to boast he could throw them at a man "and in nine cases out of ten cut out his heart."

After parting with Barnes I gathered up my "traps" and moved westward, walking alone to Beauvais' ranch at what is known as Lower California crossing, one of the places at which emigrants crossed the South Fork of the Platte river on their way to the Pacific Slope. Beauvais has one of the largest ranches on the road and the same can be said of his store. He purchases his stock of goods from time to time chiefly in St. Louis. They are taken on boats up the Missouri to St. Joseph and from there transported in wagons to their destination. His principal trade is with the Indians, with whom in fact he has been living for about twenty-five years. He has quite a number of squaws for wives—a Mormon in practice. He is reputed to be worth from $80 to $100,000. He was not at this ranch, but at the one he has about eight miles east of Fort Laramie. Beauvais is of French parentage.[59]

I staid at this place until about 4 o'clock p.m. when I started and walked to Bakers & Fales' Ranch. three miles west. Walking and carrying your bed and other traps on your back in the broiling sun is not a very pleasant past time.[60]

This country west from Plum Creek to this point and in fact to Denver is at present almost valueless for agricultural purposes. It can to some extent be utilized for grazing cattle and also sheep if it were not for the wolves. The soil is sandy and mixed with loam. It is not of itself sterile. The great drawback—the one need—is more moisture. The soil is dry and hard. It seldom ever rains on these plains and dews are not known. I have seen thunder storms in which there was a great downpour of water and have also known times when there was a day or two of drizzling rain, but these are so infrequent that, taken in connection with the hot days and the drying, scorching winds, they seem to do very little good. The valley could be irrigated with water taken from the Platte, but a system of irrigation ditches would be costly and perhaps not redeem sufficient area of the territory to justify the expense they would entail. The valleys, which alone in this part of the "great desert" constitute only a small part of the country, can only be benefited by irrigation. If this method of watering the soil ever obtains, I have no doubt the land is sufficiently fertile to produce large crops of certain kinds of grain and other products. With plenty of moisture, artificial or natural, I believe this sandy soil would be quick and active. In many places the soil is thoroughly impregnated with alkali which might be deleterious if not destructive to cultivated crops. The hot winds, which in fact are scorchers, are no friends to the farmer. It is a great danger one incurs ~~when~~ to his reputation as a prophet when he attempts to set limits on the productive powers of nature when assisted by human industry. There is no wood in this country, nor has there been any since I left Cotton Wood, which is distant about eighty-five miles. This treeless tract of country stretches westward nearly to Fremont's Orchard. There are a few trees bordering the North Fork of the Platte and also a few over on the Republican, each stream being distant from here about 25 miles.[61]

It appears to me now that these lands are the fitting abode for wild beasts and still wilder men, who now roam over them. If the government should set them apart for this purpose, reserving the right to transit over them, I do not think much would be lost to Uncle Sam. Still, waste places have been and can be reclaimed.

Had a splendid supper, which was cooked by Mr. Baker, who has become somewhat skilled in the culinary art. A good meal depends as much if not more on a sharp appetite, as on the cooking.

June 25, 1861

Staid last night at Baker's and had a good healthy sleep on a buffalo robe bed. Breakfasted on antelope meat, biscuit, and coffee. There is no meat equal to the antelope. It is sweet and juicy.

This morning Ed Creighton and my brother James overtook me. They drove up to Baker's just as I was starting on my daily tramp. They left Omaha one day after I did. They made the journey with a span of mules, Mary and Jane, and a strong but light concord buggy. I took passage with them and we were a set of "jolly good fellows."

The Platte Valley here is quite narrow, not exceeding three-fourths of a mile in width on each side of the river. For sixty miles it has been gradually becoming narrower.

Some portions of the prairie are beautiful. The prickly pear or American cacti are in bloom. The most of these blossoms are yellow and are at the top of a moderate sized oval leaf, which is covered with very sharp thorns. There are a few cacti which have red flowers. These plants frequently cover acres of ground and destroy nearly all other kinds of vegetation. I have seen five or six kinds on the prairie. They thrive best in the dry, sandy loam soil. The oval leaved cactus has only two or three leaves, the one branching from the other thus: [includes illustration]

The day before yesterday I passed through a prairie dog city. These are a very cute little animal and they are gregarious, living in towns and cities, some of which cover acres of ground. They burrow in the ground and dig their holes to a great depth. Some claim they dig until they reach water. The dirt is thrown up around the entrances to their holes to the height of eight or ten inches, forming a circle around them of a diameter of eighteen to twenty inches across the top, and sloping therefrom to the hole. In this basin they stay, frequently putting their head just above the rim and taking a survey of the surrounding city. They bark or rather make a noise somewhat similar in sound to a small mangy lap dog suffering from a severe cold. Hence I presume they came to be called prairie dogs. They look like diminutive woodchucks. They are about ten or twelve inches in length, rather heavy body and short legs. The color of their hair on sides and back is brownish red, while on their bellies it is yellowish white. It is a very difficult matter to shoot and get one. If by a chance or excellent shot you kill one, ~~they~~ he falls back and disappears in ~~their~~ his hole.

Owls and rattle-snakes inhabit the holes or homes of these dogs of the prairie. They comprise quite a happy family—which ante-dates Barnum's "Happy family."[62]

About 3 o'clock p.m. arrived at Julesburgh. Fortunately for brother J. J. he was able to find a man on his way to Denver, with whom he, at once, made arrangements to go to Denver, and within half an hour he was again on his way to his place of destination.

At this point the telegraph line is to leave the South Platte Valley, crossing the river and thence for about ___ miles running up Lodge Pole Valley and then crossing the divide, it is to follow the North Platte Valley.[63] We secured good quarters for several days' stay at this place. George Guy was here with several loads of poles from Cottonwood. After supper played cards with Guy as partner.[64]

Julesburgh is on the south side of the Platte and is about 200 miles west of Fort Kearney and 400 from Omaha. It derives its name from a Frenchman named Jules, who for years has been an Indian trader.[65] There are six buildings in this city. One store or trading house, one dwelling house or stage station, and four sheds and barns. These structures are large and commodious and well adapted to the purpose for which they are used and are built of logs. The stages for California diverge from the Denver route at this point and runs to Fort Laramie, South Pass, Salt Lake, Virginia City, Sacramento, and then to San Francisco.[66]

June 26th, 1861

Just before sun rise the stage came in from the west and its approach was announced by the blowing of a horn by the messenger and the shrill peculiar yell of the driver.[67] These stages make about ten miles per hour.

Guy left for Cottonwood with his teams for telegraph poles. It will take him about nine days to make the round trip.

Devoted the day to writing letters for Creighton. He states to me the subject matter of what I shall put into the correspondence and then I throw it into a compact readable form. Wrote three long business letters, one to Elwood, one to Wade, and one to Ellsworth, bringing matters up to date from last writing in Omaha. Wrote for myself a few brief personal and friendly letters—one to brother Randall.[68]

All day long the wind blew a gale from the south. The south fork of the Platte is about one-half mile wide at this place and the water is from one to four or five feet in depth. The wind was blowing so hard that it drove the water to the north side of the river. Only a very little was left south of the center of the stream.

June 27, 1861

Commenced making out our first financial report to the telegraph company, and was engaged in this work nearly all day.

In the morning a pilgrim had a difficulty with an Indian whom he charged with stealing his revolver. Mr. "Lo" had made tracks for the Bluffs and the man from whom the pistol had been taken, on slaughter intent, rode after him and when he overtook him he pulled a cocked revolver on him.[69] He did not frighten the Indian very much, but he was induced to return to Julesburgh. The Indian denied taking the pistol and after a while the whole matter was settled and the pistol was found in the pilgrim's wagon, where I suspect it was replaced by the wily Indian. Out of such unimportant events, as the above, by the inconsiderate acts of the whites, have arisen many of the serious and bloody conflicts with the Indians. With no special admiration for Indian character, as I have learned it on the plains, I cannot refrain from saying that these wild men have endured much and long the aggressions and wrongs, which the whites have extended to them, before they, in their way, have sought to redress them. The wonder is, not that we have had Indian wars, but that we have not had more of them.

Friday, June 28, 1861

Worked all day long in preparing report to Telegraph Co.

There is quite a large encampment of Cheyenne Indians here. They are on their way to White-Mans Fork, ~~of~~ a branch of the Republican, where the greater part of their tribe now is.[70]

Late in the afternoon there was a powerful thunderstorm. The rain fell in torrents and the wind blew a perfect hurricane. For a short time previous to the storm the hue or color of the atmosphere was remarkable. It was a pale yellowish color and resembled very much the appearance the atmosphere had at the time of the eclipse in the summer of 1854.[71]

June 29th 1861

This has been a very pleasant, cool day. The wind blowing gently.

Worked all day on the report and completed it. Handed it over to Creighton, who approved it and will send it forward to the company tomorrow.

One of Byram's trains passed today on its way to Denver. It was loaded in the main with corn and flour. There were 26 wagons, each drawn by from four to six yokes of oxen. It is a grand sight to see one of these large, well equipped trains move slowly along through the unsettled country.[72] The whites will soon possess this land; first for grazing purposes, then afterwards for tillage.

This forenoon, the Indians in camp here took their departure for the Republican river. They were in command of Spotted Horse, one of the distinguished chiefs of the Cheyenne Nation. These red men now are very friendly. "Old Spot" is a very dignified, courteous man, about fifty years of age. Fully six feet tall, powerful frame, and a good looking, intelligent native American. Just before these Indians left "Spot" came into the ranch and shook hands with all of us and gave us a friendly "good bye." Daniel Webster, the grand, or Henry Clay, the affable, could have acted the parting scene no better than Spot did it.

The Cheyenne and Sioux are and have, for a long time past, been on the most friendly terms—frequently intermarrying.

These Indians are exceedingly fond of dog meat. I saw a squaw kill one to-day. She took its life with a club. It was not a large animal, perhaps about two-thirds grown. As it was killed, so it was cooked! Such cooking would not be acceptable to a fastidious Frenchman; yet I am told it is to these people a decided luxury. Considering the Indians as he and she are, I think those words of Pope: "Lo the poor Indian," should read "Low is the poor Indian." In reference to the human races, white is the emblem of civilization. If this is true then the ranchmen and traders have done their share in the civilization of these tribes, for I see not a few half-breeds and consequently these must be half civilized!

Ragan's teams came up today from Cottonwood and brought 399 red cedar poles. These will cover sixteen miles of the telegraph line.

Today James N. Dimmock reached Julesburgh on his way to Denver with three teams loaded with bacon, flour, coffee, sugar, & c. Creighton bought his loads and hired him to work for us in constructing the line.

I forgot to mention that last night in the presence of the tented Indians and the few whites at this station I had three "rough and tumble holds" with Jack Chrissman, who is a tall, heavy built man and much stronger than myself. Notwithstanding this I managed to throw him twice out of the three bouts. The difference in our size was so marked that the result astonished not only the mixed crowd, but also Chrissman himself, while I esteemed myself most fortunate indeed. The Indians looked admiringly on me as one athletic hero.

Ed Creighton is jolly and apparently having a good time, but "like the Paddy's owl he keeps up a devil of a-thinking" on the great task before him. Ed has a great head and there is a great deal in it, and what is more, he knows how to use all there is therein.

June 30, 1861

Did not attend church today! Engaged in secular business. Had to unload Dimmock's wagons and take inventory of his stock to complete the purchase made yesterday by Creighton.

The Denver stage came in today from the west. Maj. Bradford, a secessionist, was a passenger and on his way to Missouri. He declared that the federal troops had no business in that state and her rights were being trampled under foot. This he could not stand and was returning to aid in removing the yoke of despotism placed on the necks of the citizens of that state, even if it led to bloodshed although he was a "peace man." I suggested to the major that there was at least a slight possibility he was a trifle wrong in his views, that perhaps Governor Jackson was playing the role of tyrant. He became highly incensed at my remarks. I advised him to hold his temper, continue a "peace man," and wait and see the developments of time.[73]

In the morning met George Ackley—an Omaha boy—who is now keeping ranch about 9 miles west of this place. Chrissman with fifteen wagons left for Cottonwood to get cedar poles cut by Ragan.[74]

At night saw for the first time the comet of the season. It was a degree or two above a dark cloud which lay all along the horizon in the west. The rest of the sky was clear and cloudless and the stars shone and twinkled with their usual brightness. I pointed the comet out to the boys, who declared it was nothing but a star. The tail at this time was not to be seen. I was satisfied it was no ordinary star and I watched it until about 11 o'clock. The tail constantly increased in length and width until it dropped below the horizon. It was indeed a beautiful sight.[75]

Monday, July 1, 1861

This forenoon assisted Jim Dimmock in getting his three teams across the Platte River. We hitched eleven yoke of oxen to each wagon in crossing. It is a very difficult task to cross a loaded wagon on this river at this point. The water is from one to four feet in depth and flows over fine sand which is constantly shifting. It acts like quicksand. Neither cattle nor loaded wagons can stand still in it without gradually settling down so that when a start is made, there is in reality no rest to be made until safely landed on the opposite bank. The first and second crossings of the wagons were handsomely done, but not so with the last one. The cattle had become tired and did not work well. They stalled in the center of the river and with all our yelling, whipping, and swearing we could not make them pull together so as to move the wagon. They were discouraged. Fortunately, the wagon was stopped in very shallow water otherwise the loading would have become waterlogged and damaged if not spoiled. We had to get another wagon and partially unload the one we were trying to cross, and in two hauls succeeded in getting it on the north bank. Was very wet all the forenoon, but the water was warm and the day hot.

W. B. Hibbard came up while we were crossing the river. The train he was with came in during the day and camped a short distance west of the station. The men were looking well and hearty though much bronzed by the sun. Traveling on the plains changes one amazingly and from a city gent he passes—and easy is the transition—to an ox driver, dusty, dirty, and browned.[76]

The comet is less brilliant tonight. This heavenly visitor, "wayward colt of a comet," is disappearing—running its appropriate course and to return in obedience to invariable laws.

July 2, 1861

This morning bright and early Hazard's camp was a scene of bustle and activity in making preparations for crossing his train to the north bank of the Platte River. For this purpose we had to unload and re-load a part of the freight on other wagons as the wagons were too heavily laden to attempt to drag them through the water. Some of the freight we stored, which hereafter we are to send for as we need it. While the wagons were being unloaded and reloaded and brought down to the river bank, Dave Hazard and I rode across the river in order to pick out the quickest and safest route. We crossed three and four wagons at a trip. To each wagon we attached eight or ten yoke of oxen. This is called "doubling." On the first trip we took four wagons. I piloted the way on horseback. Each string of cattle had three drivers and Hazard rode backward and forward on the off-side to keep the cattle from moving down the river. They constantly give way to the swift-flowing stream. I drove across one time. About 5 o'clock p.m. we had the 18 wagons across without encountering any mishaps—not even stalling a wagon. All of us were wet during the day.

We went into camp on the north bank. The boys were very cheerful over the good luck attending the day's work, and after supper, which consisted of boiled beans, bread, coffee, fried bacon, and apple sauce, they sang comical and sentimental songs, one of which ran something like this

Sally come up and Sally come down,
And Sally come twist your heel around,
The old man has gone down to town,
Then Sally come twist your heel around.

The Mormon boys are strong, healthy, good natured chaps and some of them possess good, fine voices. They are working their way to Salt Lake, and they have become pretty good ox-drivers. They are religiously zealous, even to fanaticism.

Commenced building telegraph line.

We commenced the construction of the telegraph line to-day. The starting point was from the office established in the station house at Julesburgh. We set fifteen poles and stretched the wire across the river. Aside from helping Ed Creighton dig the first ~~post~~ pole hole I had nothing to do with this day's work in building the line except in taking the tall spliced pole across the river and assist in setting it. The wire was carried across the river on three poles—two tall poles, one on each bank of the river and one on an island in the river. These two tall ones were made from splicing two or more together.

Slept at camp in a wagon with Milan Hunt.[77] We were badly annoyed

Worker repairing a telegraph line in 1862 or 1863. United States Army, Military Railway Service, Andrew J. Russell, photographer. Library of Congress Prints and Photographs Division

with the mosquitoes. These Platte Valley birds are devils to sing and are as bloodthirsty as starved tigers.

Wednesday, July 3, 1861

Returned this morning to the station on south side of Platte. Received letters from J. J. and R. A. [his brothers]. Wrote the latter and enclosed in letter a N.Y. draft for $1,600.00 for him to give to James Creighton (a brother of Edward) for him to use in the business of the company. This letter with draft sent to Omaha by F. Smith Esq. Received from Elwood by express $10,000.00 in N.Y. Exchange.

Hazard's men did very little to-day—digging only four miles of pole holes. Hazard's and Hibbard's men are to constitute in the main the construction force. The cattle were not used to-day. They were in excellent feed and their day off rested them up nicely.

At noon to-day George Guy came up from Cotton wood Springs with 81 poles.

Mr. Edward Creighton, whom in the future I shall mention as "Ed"—the close friendly name we all respectfully called him by and which he liked—completed all the arrangements that were necessary to be made at Julesburgh this afternoon. Leaving young Mr. Reynolds, a telegraphic operator, in charge of the office established there we in the concord buggy drawn by "Mary and Jane," crossed the river just as the sun was setting. We came very nearly upsetting two or three times on our way through the water. Once or twice the mules had to swim. We got our feet well wetted. We came up to Hazard's camp on Pole Creek and went into quarters with them. The line of wire was stretched to this creek, a distance of about 4 miles from the Platte. An additional two miles of poles were set on which the wire was not strung.

July 4, 1861

Celebrated this day in working with the construction train. Ed was anxious to be present one day and work with, watch, and instruct the men. He is an old experienced telegraph line constructor. He gave the men much sound practical advice in this character of work.

The 4th day of July! I wonder if there will be many such for this nation as it now is. Can it preserve itself from disintegration? My prayer is that it may.

July 5, 1861

Ed and I started this morning at 9 o'clock for Mud Springs distant from Julesburgh 64 miles. Our route lay up Pole Creek for a distance of 33 miles. This creek runs through a narrow valley. The roads were excellent—dry, ~~and~~ hard, and smooth. Our team was in superb condition and it cost us some labor to hold them down, at first, to a fair moderate gate [*sic*] of travel. The day was excessively hot and not a breath of air stirring. All day long did we travel and see no vegetation over five feet in height. The soil is sandy—some loam is mixed with it. The principal kind of grass is buffalo although in places there was to be found the common prairie grass, the blue joint and bunch grass. We crossed Pole Creek at what is known as the "Upper Station," at about 4 o'clock p.m.[78] From this point to Mud Springs our line of travel is nearly north and over high rolling and in places very broken ground and covers 27 miles. Not a drop of water is to be had on this route during the entire distance. The ox trains are driven over it during the night. The road is hard and smooth with many long "up and down hills." The soil is sandy-clayey mixed with loam. It is covered with bunch and buffalo grasses. These make the hilly broken ground excellent pasture lands for wild animals, especially the deer and antelope and many of these were to be seen.

The surface of the divide, as stated, is wonderfully broken. Here is to be found a broad plateau, here a narrow valley with its green grasses and a few stunted bushes; there is to be seen a cañon with its rough broken sides and occasionally a rocky cliff stands out in bold defiance of the elements. Yet the time is coming—how long will it be?—when these grounds will be occupied by the pioneer farmer. Industry and hard work will transform the present condition to one of rare production. These early settlers will undergo many hardships and privations.

About midway on this divide the stage company are engaged in sinking a broad well. It is now down to the depth of eighty feet and not an indication of any water. They are instructed to sink it to the depth of one hundred feet and then if no water signs are found, to abandon the work. It is the intention of the company, in the event of finding water, to build a station at this point. No water was found; no station erected.[79]

We reached Mud Springs about 7 o'clock p.m. The long drive and intense heat took the friskiness out of "Mary and Jane," but notwithstanding this, Mary strayed off on a tour of discovery and it took us a long time to find her.

As we expected, we found Jim Dimmock in camp here. His report on finding poles was very discouraging. Slept in the station house on the ground floor.

Notwithstanding the severe heat, Ed and I had a very enjoyable ride. In coming up Pole Creek Valley, Ed urged me very strongly to subscribe for $10,000.00 worth of the telegraph company's stock, offering to take care of the payments for me. He was confident of the investment being a very good one. As I had no money I did not feel like running in debt and so declined his kind, generous offer. As an investment it would have proved all he claimed for it.

Mud Springs Station is in fact located on the North Platte Valley at the base of the hills which mark its western boundary. It is called by this name because there are some springs here making the ground muddy and marshy. A small valley, running westerly, terminates at this place—is merged into the Platte Valley. A small clear, sweet water stream comes down the little valley.[80]

July 6th, 1861

This morning got up sometime before sunrise and "picketed out" the mules. This is sometimes called "lariatting out." The process is a simple and easy one. There is the pin, an iron rod from 14 to 20 inches in length sharpened at one end with an eye at the other, sufficiently large to take in an inch rope, as [author illustrates picket pin.] Into this eye one end of the rope is inserted and fastened and the other end thereof is made fast around the animal's neck, but in such a manner as not to slip and choke it. The rope differs in length, but is usually about 6 feet long. The pin is securely driven into the ground and the animal is allowed to graze in a circle the radius of which is the length of the rope. In picketing the animals they must be put far enough apart so that they cannot cross ropes and thus become entangled, and produce disastrous results.

After breakfast we started out to survey the cañons southwest and west of the station. Ed with a guide went off in the buggy and I accompanied Jim Dimmock with his three teams. We started before 7 a.m. After about one hour's drive we came in view of "Court House Rock." This well known land mark has two attendants. One on the north is called the "jail" and the one on the south is called the "office." They are comparatively small and unattractive. Imagination has given them the names they carry.

The "Court House," with its appurtenances stands at a considerable distance from any of the surrounding bluffs or hills. It is evidently a part of a chain or range of bluffs, the continuity of which has been broken by the action of the elements of which the wind has played a strong part in the destruction. The shape of this commanding object is conical, but very irregularly so. From the base to the summit there is a succession of steps, so that, at the distance from which I saw it to-day, it had the appearance of being terraced. It must be 500 or 600 feet in height measurement from base, and has a very bold and striking appearance. This "Court House Rock" is of a yellowish gray clay, well impregnated with lime. It has hardly become petrified. However, it is nearly as hard as white chalk and the process of disintegration is very slow. It is bare of all vegetation.

This rock, I am told, is well covered with the names of those who have visited it, with the date of their visit. Men seek fame, and the longing for immortality is great, indestructible, and manifests itself at all times and in all places. In a few years the rains and winds will efface the inscriptions on this great monument and the names of the persons who thus sought fame will have perished from the Earth. A few—only a few—have a deathless fame. Tradition and history are fame's embalmers.[81]

The day grew very warm as it advanced. At about 10 o'clock I left Dimmock's train to explore the cañons on our left and see if they contained material for telegraph poles. My search was not successful. I found only a few dead pine logs and some scraggy live trees. In prosecuting my search in the broiling sun, through ravines and over hills, I found myself soon suffering from thirst and I soon turned from looking after poles to hunting for water. In moving along I soon came to a road, but where it led to I did not know, but I followed it and yet found no water. To satisfy myself where I was, where the road led to, and where water was to be had, I ascended a high bluff and took a survey of the surrounding country, the view indeed a most enchanting one. At no great distance I could see the silver thread of a stream winding its way through the prairie. Descending I made my way to the rivulet and there I quenched my thirst. Then I went up to a camp and there found a brother-in-law of Judge Eleazer Wakeley of Omaha.[82] On my way up into the cañons where Jim Dimmock was in camp I saw a herd of antelope and their fright at seeing me sent them scurrying over the plains at full speed. Found Dimmock and his men eating dinner. I was invited to take "pot luck" with them and gladly accepted the invitation. The food was coffee, bacon, beans, and bread and with the keen appetite I had, I can safely say no king ever enjoyed his kingly repast more than I did this frugal wholesome dinner.

In the afternoon worked with the men in getting out poles. This was very slow and hard work. The "poles" were crooked, [k]notty and most of them good sized trees, dead, dry, and brittle. We cut and piled 18. Getting <u>such</u> poles was very unsatisfactory, but we had to have them. At night we went down and camped on the little stream. Slept in the wagon.

Sunday, July 7, 1861

This was indeed a very fine morning. The sun lifted himself up out of the plains looking red and angry, presaging a very hot day. Dimmock and men took a rest. After breakfast I followed the stream downward and found some wild currant bushes with berries large, ripe, and decidedly luscious. I discovered some gooseberry and black currant bushes. There was only a small quantity of fruit on them. There was an abundance of wild choke cherries, not yet ripe. I am told the Cheyenne

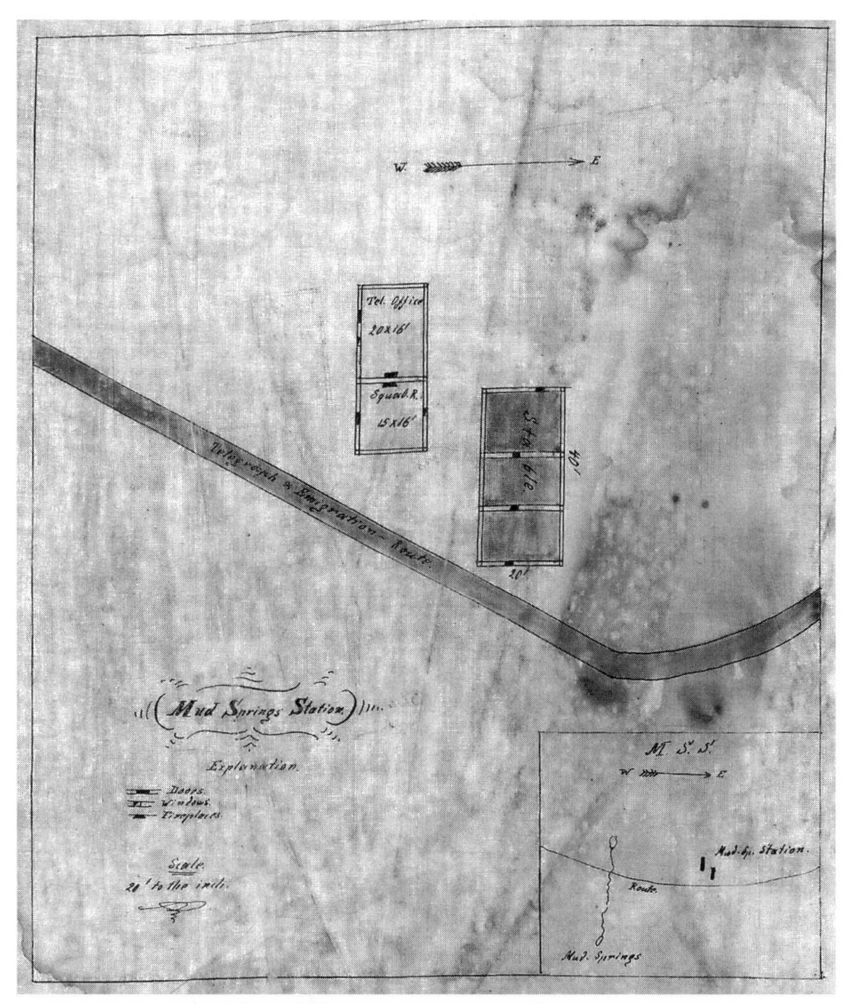

Map drawing of Mud Springs Station by Lt. Caspar Wever Collins. Courtesy of Archives and Special Collections, Colorado State University

Indians gather and dry them for winter use. I disbelieve this, as they are such an improvident set of people.

In the afternoon took the ox-express for Mud Springs. Young Comstock was the driver. He selected the sprightliest and wildest yoke of steers he could find and hitched them to the express van which consisted only of the fore-wheels of the wagon. With difficulty we kept them in the direction of the Springs. They ran away with us, up and down the hills, through gullies and valleys, and over rocks. Several times we came very nearly overturning. We put the whip to them and let them slide. It took us a little over one hour to make the six miles.

At the Springs I found Ed Creighton and the train under [the] Hazards. Took supper with [the] Hazards. About dark [George] Guy and Milan Hunt and men came up. They are to get out poles on Lawrence Fork of the North Platte River.[83]

Yesterday Ed went out hunting poles with Chris, the stage company's hunter. They had no success in hunting poles, but did well in hunting antelope. They killed and brought in four of these animals.

During the afternoon Mr. Benham came to the ranch. He is one of the agents of the overland stage company and he is a very pleasant intelligent gentleman.[84]

Ed leaves tonight for Julesburgh in the stage to be absent for a few days.

July 8, 1861

The stage coaches are now running daily and with considerable regularity to the west. The[y] passed through or rather by Mud Springs on the 1st as a daily line, carrying mail and passengers. Creighton has made arrangements for me to ride on them free of charge, when I wish to. Took stage about noon for Lawrence Fork, about ten miles from Mud Springs. Reached the fork about 1 o'clock and found the Hazards in camp. Took grub with them. They are on their way to Chimney Rock, from which place they will look up poles to cut and draw on the line which in the main follows the stage route. My object in coming to this point was for the purpose of marking off the ground between this point and Mud Springs for distribution of the poles at or near equal distances apart. We allowed from 22 to 25 poles per mile. If ground was level 20 good poles were made to do the wire supporting. For the purpose of aiding the wagons in unloading the poles at the right place, I started at a point on the Fork and measured off 70 to 73 paces and then dug a small hole and made a mound of ~~with~~ the earth excavated with a spade I carried. These mounds I made of sufficient size to attract the attention of the person on the wagon distributing the poles. About sun down brought up at the springs pretty well worn out. The staking out or mounding out of ten miles was a big afternoon's work. Took supper then played cards, and then went to bed.

July 9, 1861

This is a glorious morning, but bids fair for giving us a hot day. The day is mine to do with as I please, and as the new work yesterday—digging holes—called into action a new set of muscles, and I over worked them. I am now suffering from their soreness and stiffness. For violated law there is punishment.

Yesterday Dimmock, in the afternoon, brought in 32 poles and logs, scraggy, crooked, unsightly things. Yet they must be utilized.

In the forenoon did some work in peeling Dimmock's green pine poles.

Took a good long snooze after dinner on the ground in the shade of the log and adobe station house.

Hunter Chris failed to bring in any game to-day. He managed to get only one shot to-day and that was at an antelope and singularly enough he missed the mark and the game went bounding away. He is accounted the best shot with a rifle of all the hunters on the line of the road. When he starts out he takes two good strong ponies, one for riding and the other for loading with his game. The deer and antelope, in their curiosity, surpass the female portion of the genus homo. They are exceedingly timorous, and fly from danger, but like Lot's wife to their destruction they cannot resist their inquisitive inclination to turn and look back and see what it was that gave them the fright. This means death to one or more of them. Chris when hunting wore no hat, but bound around his head a white or red cloth. If, when he saw his game, he could not steal up to them and get a shot, but frightened them away, he would get behind a rock or some sheltering object or failing this he would throw himself on the ground and lay quiet. After the antelope had run beyond danger, it would stop and look back and around to see what it was that gave it the fright. Then Chris would get in his "good work." From behind his hiding place or from the ground he would raise his head and move it around. The animal would soon have its attention attracted to this waving object and irresistibly back it would come to investigate. When within easy rifle range Chris would drop him. After supper went out and marked for planting poles on the line southward from this place. Laid out two miles and then walked back to the springs after sun down. The day has been very warm.

The North Platte River is about six miles from here, one of the widest places in the valley. The soil is chiefly sand with small admixture of loam.[85]

The nearest point from the springs at which wood can be obtained is six miles west and southwest in the cañons and hills. The chief kind is pitch pine which for fire purposes can be obtained in considerable quantities. A few box elders and cottonwoods are to be found. These hills stretch to the west rising the one above another, until in the far distance they assume the appearance of small mountain peaks. The soil is sandy, dry and suitable now only for grazing. The snow does not fall in sufficient quantities to prevent cattle grazing and doing well during the winter months.

There is sensibly a difference in the weight of the atmosphere here on these plains and in these hills and what it is at Omaha. It is much rarefied and is free from all moisture at nearly all times. In breathing the respirations are much quicker than at the ocean's shore. This place must have an altitude of ________ feet above that of Boston. No dews fall here.[86]

Mud Springs Station is owned by the stage company and is kept for it by a man of the name of Kent. He and his wife have been worshippers at Brigham's shrine. They had abandoned the faith, or rather deserted the brethren and kept the faith. They spoke in severe terms of those who managed the affairs of Utah and declared that the condition of the women was no better than that of the slaves of the South. This may be so. Of one thing I am certain and that is that plural marriages or polygamy must have a degrading tendency.

July 10, 1861

Early this morning took the stage and rode to the camp on the divide where the men are digging a well for station purposes. This is about 12 miles south of Mud Springs. Walked back and marked holes for distributors of poles. Returned about 3 p.m. George Guy came in with his wagons loaded with poles. He had fifty-five. These, notwithstanding some of them were scarcely suitable, we distributed on the line northerly. He returned to the cañons for another load, which he will bring in tomorrow.

Heard from Hazard's train. He was in camp near Chimney Rock and unloading his wagons to use in handling poles.[87] This rock is distant from this place twenty-five miles. It is named "Chimney Rock" because viewed at great distance it resembles the chimney rising out of an eastern dwelling house ruin.

The route I have traveled thus far on my way into the mountains, taken as a whole, is a vast inclined plane and I should judge its ascent from the Missouri River to average four feet to the mile. Thus far a line of rail road could easily be constructed from Omaha upward to this place, by following the ~~South~~ Platte River to ~~its~~ the confluence of the North Fork with it and thence up the valley of this last stream. The grading of the road would be a work of no difficulty. The number of streams to be bridged, including draws from the hills which sometimes are bank full, are comparatively few and of no great width or depth of water. The large streams are three if the north branch of the Platte is crossed to the west and south side: to wit: Elkhorn, Loup Fork and North Platte. The stone used in bridging these three principal streams would have to be quarried in and brought from the states east of Nebraska, as in the country where the road would cross them there is no stone suitable for such uses. The procuring of ties will require an unusual expense in this rail road construction, as they must be taken from the pine woods in Michigan or Wisconsin or hauled by wagon from the Rocky Mountains.

It is said that coal, iron, and most kinds of base metal as well as silver and gold are to be found in the mountains west all along the Rocky Range. As for me I have no doubt of this. The mining of these will give rise to great industries, and their manufactured products must find a market in the East.

A rail road once built to the golden gates of San Francisco connecting the two oceans by a cheap and expeditious method of conveyance would, as it now seems to me, be a good paying investment on the money required to construct the road.

It now costs two hundred dollars to travel from the Missouri River by stage coach to San Francisco. The travel by coach is very limited owing to the hardships and dangers of a twenty day's ride through a country uncultivated, rough and mountainous, and uninhabited only by wild, savage, and unreliable Indians. At present they behave themselves, but when they will be on the rampage, become intractable, commence their depredations—plunder and murder—no one can foresee. There has been for quite a number of years, and it is likely to continue for many to come, a vast migration from the States to the territories west of the Missouri and the Pacific Slope country. Not only would a rail-road take almost all of this travel but it would induce a far greater movement to these inviting fields. With a road how quickly these now unpeopled lands would be possessed by an intelligent, industrious and enterprising race of men! How soon then the desert and wilderness would "bud and blossom as the rose" and great states leap into existence and exert a powerful influence in our national government! With or without a rail-road the "star of empire moves westward." Development, production, commerce follow civilization. Why then do not some of our monied men of the country and especially of the East invest in this great national enterprise. A continuous line of road from San Francisco to New York would be a highway for nations to use in travel and transportation of their surplus products. What a vast amount of inter state and inter national commence would find its way across our country. The work when undertaken will be stupendous, requiring years to complete and the expenditure of millions of dollars; yet with the assistance the government will furnish and with the energy and enterprise of our men of wealth, I can see no good reason why this need in proper time cannot be supplied. How many years will intervene [between] the present and the completion of a rail road across the continent remains to be seen. That it will be built I have not the least doubt. Will we have to wait long? "The world moves."[88]

To-day saw a fine flock of antelope and when they saw me how they went bounding away over the prairie. I also took note of two new kinds of cacti. The one was in bloom and the other was over its blossoming period. The one consisted of a pulpy mass covered with small sharp prickles. Take a large pine cone and have thorns growing from it thick and from every part of it, and it would resemble this variety of cacti. It was in bloom and its flower was at the apex of the cone, small and vermillion in color. The other kind was much smaller and many of them grew from one parent stock. They branched out from each other. The leaf of this cactus is round, about one half inch in diameter, and from one to two inches in length. It is also covered with prickers.

We are now setting down into good hard work and the building of the line will progress from this time forward rapidly. The trains of Guy and Dimmock are at work in the bluffs and cañons on Pumpkin Creek about 8 miles southwest of here. Hazard's are over near Chimney Rock on the Lawrence Fork. Hibbard has charge of the construction squad of our men and is now working his way up Pole Creek. Ragan and Chrisman's trains are hauling poles from Cottonwood and Julesburgh and distributing through Pole Creek Valley. Joe Creighton's train is somewhere east of Julesburgh and I guess spends most of its time in camp. Joe is a good hearted fellow, but was never born to build telegraph lines on the prairies. Jim Creighton is pushing to the west with rapidity and will commence cutting and distributing poles somewhere in the hills west of Ft. Laramie. John A. Creighton with a small train is coming on from

Denver to join us. Soon all the different divisions of our entire force will be at work on the construction of the line. Mr. Creighton estimates that we must average at least eight miles of constructed line every day in the week and every day in the month. Counting the days we cannot work, to realize his estimate we must build each working day ten to twelve miles of the line. This I believe we will be able to do—as the men in charge of the trains are pushers—except Joe Creighton.

Thursday, July 11, 1861

This day until nearly sun down, has been unusually hot.

Guy and Milan Hunt came in today with two loads of poles, 27, making in all they have cut and distributed to this point eighty-two. Assisted him [them] in distributing them. They returned to the cañons to cut more.

Dimmock came in with 48 poles. This makes the number he had finished eighty. He has 25 cut in the bluffs.

The thunderstorm was terrific. The rain fell in torrents, the wind was high, the flashes of the lightning sharp and almost incessant, and the thunder deafening.

I expected that Ed would come up last night from Julesburgh, but he failed to put in an appearance.

The comet is yet visible.

July 12, 1861

This morning the temperature was in marked contrast with that of yesterday.

In the forenoon aided Dimmock in distributing poles and returned about 12 o'clock noon. Did some tailoring—mending pants and shirts. This trade, which I learned at preparatory schools and college, I find of great advantage while "roughing it on the plains."

After dinner helped Mrs. Kent pick over some wild currants, which one of the men had picked out on Pumpkin Creek. Some of them were ripe, exceedingly palatable, and tasted very much like the ~~wild~~ eastern black currant. These currants were black and yellow in color. I selected some of the ripest and took the seed from them to send east.

Mrs. Kent and I had a long friendly chat on various matters, and finally drifted into a discussion of Mormonism. She and her husband were "back-sliders," although from her conversation I soon learned that the close communion Baptist doctrine, "once in grace always in grace" applied with force to her. She denounced, in strong terms, the practices of "the Saints" in Utah, but not a word would she say against their religion. The religion was all right, but it was not lived up to—was abused. How strange it is that we find people standing up and vindicating a system of religion which has made them miserable! The very error of the system is the poison which works the misery. She told me all about her exodus from Utah. Her stay in that territory had been an enforced one for over one year and she and her husband stole themselves out of Brigham's realm.

The daily mail coaches from St. Joseph, Missouri, to Sacramento now pass here every day. The first daily stage from California has not yet passed here. It is expected to pass this point eastward during the day, and afterwards every day there will be a daily mail each way.[89]

Creighton did not come up on the stage.

Late in the afternoon there was quite a hard thunder storm. As the storm in its severity passed it gave us one of the most beautiful rainbows I ever saw. The band was broad and the colors brilliant. It extended from horizon to horizon. Wrote Lewis.

July 13, 1861

As a diversion went out and worked during the forenoon in digging telegraph pole holes. Returned to the station for dinner and found Guy on hand with 32 poles. These we distributed on the line north.

First east bound daily coach.

A short time before sun down the daily coach from California arrived at this station. It left San Francisco on the 1st inst. It had on board four through passengers. The fare is two hundred dollars from St. Joe to the

Golden Gate. The schedule time between these two places is for winter, twenty-three days, and the remaining part of the year twenty days. This first coach will make the distance in about 17 ½ days. It being the first east bound stage, the messengers are trying to beat schedule time. The coaches are the old Concord stage style. There are four horses or mules to each one. They came to this station on a run. The messenger blew his horn. The driver cracked his whip and gave his unearthly yell of yip! yip! yip![90] A change of team was quickly made. The messenger shouted, "All aboard." The passengers scrambled into the coach. The horn gave forth its music, the driver yelled and swung his whip, and away went the coach in a cloud of dust.

The company receives from the United States one million of dollars for carrying the mail each year. There is money in this contract.[91]

The sun sets at this point are gorgeous. I never in my life saw one so matchlessly beautiful as that of this evening.

"The sun descending set the clouds on fire."[92]

Heard from Creighton to-day. He is delayed at Julesburgh owing to the non-arrival of Chrysman's [*sic*, Chrissman] and Thompson's teams.[93]

Joe Creighton has crossed the North Platte and is on his way to this place.

Hibbard and his construction company are camped at the last crossing of Pole Creek to which point they have the holes dug and are now waiting for the poles, which Creighton is trying to push forward. Undoubtedly Hibbard will continue digging holes.

Monday, July 14, 1861

Spent this warm, delightful day in writing letters, playing cards, and sleeping. Surely there is no Sunday on the plains!

July 15, 1861

Started this morning with Cy and Chris, the hunter for Dimmock and Guy's camp, to have them load and distribute the poles they had cut and then move forward and find a new place from which to cut poles. As my companions had ponies to ride and I had none I had to walk and so started in advance of them. After about an hour's walk I discovered an antelope grazing not far away. I concluded to try and shoot him with my pistol. Putting a large rock between me and the harmless animal, I began to skulk and sneak upon it. Unseen I walked upon a bed of prickly pears or cacti, and filled my moccasined feet full of their thorns. I forgot all about the antelope and with a bound, and some vigorous exclamations, I was soon off the dangerous ground. Of course the antelope saw me and quickly made his escape, perhaps, however, not from danger from my revolver. I removed the briars from my feet and moccasins the best I could. When I narrated to Chris my hunting adventure, he enjoyed a good hearty laugh at my expense.

When I came to D[immock] and G[uy]'s camp I found them making preparations to desert it for good, and assisted them in "breaking camp." Returned with them to the stage road and took dinner where it crosses Lawrence Fork. The poles reached three or four miles north from this stream of water. Returned to Mud Springs Station, which I reached at 10 o'clock night.

The east bound stage, which came in that night brought the startling news that one of the Pacific Telegraph Company's men had been drowned at Fort Laramie, while attempting to cross the Platte river. From the best information I could get out of the stage driver I was led to believe the drowned man was either Jim Creighton or John McCreary. I went to bed—which was on a buffalo robe spread on the ground—with this impression on my mind. I obtained very little sleep, as can well be imagined under the circumstances.

July 16, 1861

Ed Creighton arrived at this station on the coach at 2 o'clock a.m. I immediately got up and informed him of what I had heard about a man in Creighton's train being drowned and the reason of my fears that the man was either J. C. or J. McC. After consulting about the matter it was deemed best for me to take the outgoing stage and go forward and learn the true state of the matter, and if needed render such assistance as I might be able under the circumstances. Accordingly I took passage in the coach.

In the morning took breakfast at Chimney Rock Station. The mosquitoes hung around and over the station in great swarms. In all my experiences on the plains I have never seen these bloody pests in such great numbers. To protect the stage passengers from them during the morning meal, there was a big smudge of fire at the door-way, and in the dining room under the table two smaller ones. The smoke in the room was dense and almost suffocating, but still the blood seekers got in their work. Of course these precautions adopted against them were comparatively effectual. One can judge that these pestering insects were very thick when I state the fact that during the night they stampeded a small body of Indians, who were in camp near the station. They were driven away and pulling up their tepes they moved over on an island in the North Platte River.

This station is named after the high tower rock some twelve or more

An 1866 ink drawing by William Henry Jackson shows a wagon train passing Chimney Rock, with telegraph poles visible beside the trail. Courtesy Scotts Bluff National Monument.

miles to the west. It is a clay tower, remaining in a chain of bluffs or hills from the Platte river running south westerly. Three bold headlands of the same remain to wit, Court House rock, Chimney rock and Scotts Bluffs. The clay of which this monument is chiefly composed is nearly petrified and almost impervious to the attack of the elements.[94]

It is very difficult for one to measure with the eye—guess—with accuracy the distance he is from any large object in the distance. Chimney rock would not be adjudged, by the ordinary pilgrim, to be more than three or four miles from the station after which it was named, when in fact it is distant a large twelve miles.

One of Hazard's men, "French Pete," said he could walk to this rock from their camp near the stage station and return while they were getting breakfast and yoking up the cattle ready for a start. This was of course disputed and Pete started out to convince "the boys" he was capable of doing the walk. Breakfast was cooked and eaten and Pete could be seen marching toward the rock at a swinging gate [*sic*] The cattle were yoked and hitched to the wagons and the train moved out and still plucky Pete was moving forward with his face set to the rock. Pete said he went to his point of destination, but he did not join the train until late in the afternoon. He had tramped nearly all day long without anything to eat. He was weary, foot sore, hungry, and completely played out and never afterwards could be induced to guess on distances on the plains.[95]

After eating and allowing the mosquitoes to eat, we rolled away from this station on a running gate [*sic*]. The roads in the main are splendid—smooth and hard. Occasionally we have to cross a sand-bar or the bed of a dry creek. Here the wheeling is heavy owing to the loose, dry sand. The distances across these dry creeks is short. The stage stations are about twelve to sixteen miles apart. At every station the animals are changed. We passed through Scotts Bluffs, thence up the valley of the North Platte and on the south side thereof. We passed Beauvais' upper ranch about sundown and reached Fort Laramie just after dark. Called at the Fort's post office for letters but found none. The stage coach made a stop at this place for about one hour and then the mules at a rattling pace drew us westward into the black hills. The ride through and over these hills, in the night time, was far from pleasant.[96]

This fort is nicely located in a rise of ground west of Laramie river and south of the North Platte. I judge its capacity to be three companies. It could easily and most comfortably accommodate them. This fort was erected as a protection to the early pioneers crossing the plains and as "a moral restraint" to the roving bands of Indians in their murderous attacks and depredations upon the restless and migrating whites seeking homes on the Pacific Slope. I was at this Fort several times after this and found the grounds and buildings in a cleanly condition. The houses were constructed of logs in the main, although some were of adobe material. The officers and soldiers were gentlemanly in their deportment.[97]

July 17, 1861

At sunrise came up to Jim Creighton's camp and found I was mistaken in my opinion of the man who was drowned, for Creighton was still in flesh and endowed with power of locomotion. The man who was drowned was named Thomas White. The place of this accidental death was north of Fort Laramie. He had crossed to the south bank of the Platte on mule back with one or two other of the men to see if they could get any mail at the Fort. In re-crossing the mule "rolled" and White was thrown into the river—then high and rapid—and was unable to save himself. Creighton lost another of his men on his westward trip near Plum Creek. This man was named Wells and was a resident of Omaha with whom I had a passing acquaintance. In the language of Jeems Hambrick: "He was accidentally shot." With the two above mentioned fatal cases, none of the men connected with the outfit have had any suffering from sickness.[98]

After breakfast the train moved on and I climbed into a wagon and "slept the sleep of the righteous" until noon, camping at Horse-Shoe Creek. Refreshed and hungry, dinner was truly a luxury.[99]

At this point Creighton commenced to cut and haul poles for the line.

In the afternoon we unloaded all the wagons, except the two "grub wagons," even taking off their beds. After unloading went down and took a good bath in the creek. After supper played cards, Jim Creighton and O'Neil against John McCreary and myself.

July 18, Thursday.

Started early this morning with the whole train for the foot-hills of Laramie Peak to cut telegraph poles. Our direction was a trifle west of south. The train wended its way up the creek keeping in the valley thereof. I left the train about 9 o'clock and struck out into the hills as an exploring party. On these hills and in their ravines I saw for the first time since leaving New York, the New England red breasted robin, the blue bird, and yellow bird. It made me cheerful to see again these old familiar friends. I found in these ravines deer ~~horns~~ and elk horns. There was one pair of elk antlers of most stupendous size. The train made only about eight miles before camping for dinner and cattle rest.

After dinner Creighton and McCreary started on mule back to select a place to cut poles, while Hengen, whom the boys called St. Louis, and I with guns and revolvers started out to see if we could bring down any game and as well to look for poles. We found in one place an abundance of straight sizable young pines suitable for our purpose, but from the lay of the land about them very difficult to haul them out. We discovered wild currant, rich and luscious, and feasted on them. "St. Louis" and I finding no game amused ourselves in shooting with rifle and pistol at a mark. Returning to camp, rather late, I was not a little astonished to find there Ed Creighton, who had followed after me to know the truth of the reported drowning of Jim. He came up with a span of mules and buggy which he obtained from Beauvais, a French trader located a few miles east of Fort Laramie. Ed is a steam engine of energy and has wonderful powers of endurance and the enterprise he now has in hand virtuously compels him to be ubiquitous. J. C. and J. McC. did not come into camp until very late. They found poles in abundance at the base of Laramie Peak and easy of access. After supper as usual there was a social game of cards. The game was "Old Sledge" or "seven up" or "high, low, Jack and the game." If it were not for cards what would the poor devils on the plains do for amusement?

July 19th, 1861

At a very early hour we were roused from our slumbers with the shrill voice of Creighton of "Roll out"! Roll out!! and immediately the encampment was in a bustle. By sunrise the train was moving towards Laramie Peak. I was assigned to drive the "cavy yard" or the loose cattle. When near the place of encampment for noon George Roth and I had some amusement in killing ducks in the creek. We killed several but only secured two. In the afternoon all the men were busy in cutting,

Laramie Peak, present-day Wyoming. NSHS RG2955-47

peeling, and hauling out the poles for loading. There were only two excellent wood choppers in this whole camp. They all however pitched in and did good work. At this season of the year the bark would not "strip" from the young trees and the "peeling" was done with draw-shaves which we took the precaution to bring with us. This camp could appropriately be called "Camp Industry."

On our way up to this peak yesterday we saw what are called "horned toads." They look very much like the common toad, but not so large, and yet in some respects it has a striking resemblance to a turtle. The head is like the common toad. Just back of it is a row of stubby horns. The legs of this reptile are like those of the turtle. It has a tail like a lizard. Those I saw were sluggish in their movements. They are found on sand banks and where there is no water. I presume they subsist on flies and other insects. What in the economy of nature they were created for, even a live Yankee could not guess.

The rattle snakes were plentiful around "Camp Industry" and I walked over one this afternoon and when it sounded the alarm, kindly gave warning by his rattle of the danger I was in, I, unconsciously, did some tall levanting [*sic*, levitating]. The very first thing which a man does when he hears this snake's ominous rattle is to jump whether to or from his snakeship.

Saturday, July 20, 1861

Called up this morning at light and directed to go in search of the mules, which had pulled up their picket or lariat pins and "gone off." Saddling McCreary's mule I started down the creek. We came up and came upon them about four miles from the camp. After chasing them about three miles I finally succeeded in driving them into a wild plum thicket and took them prisoners and returned in triumph to the camp. I had a good breakfast; hunger was the sauce. Remained in camp until Ed came down from the woods and informed me we must return to the several camps near Scotts Bluffs and push the work that was being prosecuted there.

From a cursory observation I judge that round about Laramie Peak, the grounds are most excellently adapted to spring, summer, and fall pasturage; and I am told by old French trappers and ranchmen that seldom in the winters the snow falls to any great depth or remains long on the ground, which holds no large amount of water. The atmosphere is dry and invigorating. Seldom are there any dew falls. Small streams of very pure water are found in good numbers at the foot hills of the Peak. These streams run only a short distance after leaving the hills and gradually sink into their sandy bottoms and disappear. The trees and brush scattered on the hills and in the ravines afford excellent protection to the grazing animals from the occasional heavy storms. There is to be found some wild prairie grass in the valleys. The long stem blue joint, which is an excellent grass, is here to be found, but the best of all wild grasses is the bunch grass, growing here in abundance. It is very nutritious and, as a consequence, a little of it daily not only supplies the needs of the grazing animal, but rapidly fattens him. It is an early grass and has matured and ripened by the middle of September. It stands as cured hay and the late fall storms affect it little—do not wash out of it its food strength. It apparently does not seed, but in reality it does. Its seed is just above the ground and the cattle cropping it closely get them. Should this section prove what it seems to me from observation and facts otherwise obtained, there will in time be thousands of cattle on every hill and the land will, in this regard, be a source of wealth. The waste places shall yet minister to man's needs and pleasure.

There is another kind of grass which grows only in the valleys, and there it is found generally growing in the old buffalo wallows. It seems to have an affinity for alkali. Its height is not over four inches. The color of the blades is bluish green and they end in a sharp thorn. Neither our oxen, mules, nor horses would eat this grass and I doubt whether any animal does. We sometimes called it "salt grass" and also it was known as "needle grass."

There was also another grass with fine curly leaves which made a mat on the ground so thickly did it grow up. Its height was not over one inch.

It did not grow tall enough for the cattle to crop it.

The trees in this section were the pine, the box elder, cottonwood and elm, and quaking asp or poplar. Most of the streams were bordered with the willow; clumps of wild choke cherry was frequently seen. There was to be seen a bush from three to ten feet in height bearing berries evidently of the same kind but differing in color, red and yellow. These we called the "buffalo berry." Sage brush and grease wood covered the ground, and attain a larger growth than on the Platte Valley.

The soil is not sterile like a sand waste and under required favorable conditions all its products would thrive most splendidly. It is composed of sand, clay and vegetable mould. It has been baked in the sun for centuries and the wonder is, how, with such a meager annual rain fall and at such an altitude, any vegetation can grow.

There are not many wild flowers to be seen either in the valleys, ravines, hill sides, or on the mountain ranges. The sunflower, of stunted growth, is found all along the sides of the old emigrant road. The wild rose of the Rockies is just as beautiful as though it were growing in eastern Nebraska. Most of the flowers I saw on the trip were unknown to me.

Most of the days of this month have been filled with sunshine, although we have been visited with several thunder storms attended with high winds. The lightning at time was terrific and the answering thunder would indicate that it had split the sky wide open. The heat during the days has been very great, but by ten o'clock at night it had so fallen that it was comfortable and the coolness banished the mosquitoes and invited sleep.

The road up the North Platte to this place is most excellent. Of course from Fort Laramie westward it makes its course over the Black Hills, yet, with few exceptions, no steep grades are encountered.

The nights, when clear and the moon is at its full, are wondrously luminous and in this night light I have read fair sized type printing. The rare atmosphere, free from moisture, must conduce to this condition of light.

Returning from this digression to the narrative in hand Mr. E. Creighton and I started to return to the trains that were employed in the region of Scotts Bluffs. We took the mules and buggy which brought Creighton to the camp. When we reach[ed] the stage station, we had to lay by for about two hours on account of a severe storm. We left Horse-Shoe Creek station about sundown intending to reach Bitter Cottonwood for the night's camp. Owing to the darkness the mules took a camp road. This caused the loss of about an hour's time. A short time after we regained the mail or true road, the stage coach was driven up to us and Ed stopping it, took passage to the east, leaving me in the darkness and on an unknown road to find my way to my objective point for that night's rest. It was late when I reached my destination and rousing the man in charge I made known my wants. I inquired where I would find grass for my animals and he told me there was none on that creek. Securing them for the night I sought my assigned quarters in an Indian tepe and soon was oblivious to what was going on in this world. In the morning I found the station was in a distressful poor section of the country. It is located on the bank of a very muddy stream which is bordered with cotton wood trees.[100]

July 21, 1861

Settling for my night's entertainment I harnessed my team and put out for some place where I could get "refreshments for man and beast" and drove up to Center Star ranch at 8 o'clock a.m. This place is kept by Jules Cuny [*sic*].[101] Gave the mules hay and corn and made a good breakfast on bread and milk and coffee. To me this was a most sumptuous repast. By 9 o'clock I was rolling eastward. During the forenoon I had what I perhaps manufactured into an adventure. When I had reached the summit of quite a long hill, an Indian, who had been laying by the side of the road, rose up and when I came alongside of him he said in fairly good English: "Give me a ride." I said no. He was armed with a gun. He kept by the side of the buggy and every few rods repeated his request: "Give me a ride." I started the mules into a good trot and yet this Indian kept by the side of the buggy and repeated: "Give me a ride." It appeared to me that this was growing monotonous if not likely to be dangerous. Acting on the spur of the moment I wound the lines around my leg and crossing the other over it to prevent the lines from unwinding I grabbed up my double barreled gun, cocked it and kept it pointed at this noble red man. Holding "the drop" on him I urged the mules into a good pace. Mr. Indian was evidently content to drop behind and perhaps realized that a frequent iteration of "Give me a ride" had lost the pith of a joke. Of course while holding the drop on the Indian I had paid no attention to the mules. They however kept the road but had settled down into a run. It took me some time to get them under control but the Indian was distanced in the race. It is a great wonder to me even at this remote day which one, the savage or civilized man, was the largest sized fool. However I took no chances and rejoiced when the adventure was over. I am in hopes that Indian had a long life and filled it with glorious and useful deeds.

Reached Beauvais' about 1 o'clock p.m. Returned Beauvais his mules and took "Kit and Mary" in their place and continued my journey to the camps below. Came to Horse Creek ranch, so called from its location on a stream by that name, at sun set, and concluded to stay all night. A man by the name of Reynolds [*sic,* Reynal] keeps this ranch and stage station.[102] He has sold out his interest in his "shebang" to a man by the name of Bordeaux and is to leave in a few days for a place on the Arkansas River.[103] He has a large herd ~~of cattle~~ of mixed cattle, from a Texan steer to a short horned Durham bull. They were all fat and sleek, and killed and dressed would adorn a butcher's stall in Fulton Market and well cooked would satisfy the taste of an epicurean when eaten by him. They were grass fattened and were a good advertisement to the nourishing qualities of the grasses upon which they had fed.

July 22, 1861

At eight o'clock was on my way to Scotts Bluff, through which I passed and reached Lurjon's ranch, eight miles east, at three p.m. This was a sweltering hot day. During the after-noon H. Powell, one of Joe Creighton's men, came up to this ranch with two loads of wire and insulators and I helped him unload and store them away. From him I learned that the wire had been strung as far as Chimney Rock.

July 23, 1861

Wrote several letters for Ed and one for myself. After this paced off the ground between Lurjon's and Scotts Bluff and marked the places for distribution of the poles. Suffered very greatly from the sultry heat and thirst. When I reached the broken grounds around the base of the Bluffs I made search for pools of water in the hard clayey earth of which in the main the Bluff is composed. Was successful in finding what I sought for, but it was muddy, very warm and nasty, besides the mud it held in solution it floated not a few dead crickets and grasshoppers. Brushing these away I lay down and drank the muddy compound in the manner I used to drink the pure waters in the eastern New York streams. Nevertheless it slaked somewhat my thirst. I concluded I would wait for the stage and ride it to Lurjon's rather than to walk back. Finding a place to lie down in the shade and wait for the coach, I concluded to stretch out and "take things easy." My resting place was close to the road. Naturally I fell to sleep. In a short time the stage came along and when the lead mules saw me they gave a big shy—a jump sideways—from me, and when I regained consciousness, George the driver with whom I was acquainted was busily engaged in swearing, in holding in his team, and holding a cocked pistol on me. My position for a brief time was rather precarious, but by throwing up my hands to show no harm was intended he soon regained his coolness. I approached the coach and told him I wanted to ride down to the next ranch. He invited me to take the seat beside him and soon I explained the whole transaction, so far as I was concerned, to him. He said his first impression was that he was being attacked by Indians or stage robbers. We had a good hearty laugh over the affair as we rode along.

At night Hazard's train came from the hills near Court House Rock, laden with poles for distribution. They were cedar and pine. The train went into encampment a few rods west of the ranch. Ed Creighton a little after dark came up from the camps below and he and I slept together, at Hazard's camp, under a wagon.

July 24, 1861, Wednesday

I accompanied [the] Hazards in their distribution of poles westward. I found that my marking the places for the poles greatly facilitated their work. Made about twenty miles but had to drive late. From Scotts Bluff to where we went into camp I marked off the ground.

July 25, 1861

It was late in the morning when I commenced marking the ground westward, for distribution of poles, to Horse Creek. Here we found Ed Creighton holding conversation with Hooper, a delegate in Congress from Utah. He was journeying homeward with four or five wagons each drawn by four mules. He had with him several good looking women.[104] I was introduced to Mr. Hooper but not to his women. Reynolds [*sic*] had surrendered possession of the ranch to Bordeaux and had departed for Arkansas. Towards night went along with [the] Hazards into the hills. We camped near a slough when there was an abundance of water such as it was. Creighton stopped at the ranch.

July 26

Mr. Creighton came into camp with our mule team bright and early, while we were taking our breakfast. He considered it a great joke on [the] Hazards. At noon we camped on a slight rise of the ground, just above a clear, cool-running stream of water and about 2 miles from the grove of pine poles we were to cut. After writing some letters for Creighton on business matters, took an ax and went and helped the men chop poles. Ed staid with us during the day and night.

July Saturday 27, 1861

Worked all day long with the men in the pine groves in cutting and peeling poles.

July 28, 1861

We cut a few poles this morning to round out the loads and used the rest of the forenoon in loading them on the wagons. The train in the afternoon rolled down to Horse Creek Station. Here we stopped only long enough to unload four wagons and then moved on down towards Scotts Bluffs, going into camp about seven miles below the station.

A watercolor by William Henry Jackson shows both Chimney Rock and Scotts Bluff on the horizon. Courtesy Scotts Bluff National Monument

July 29, 1861

Commenced an early day's work and from last night's camp we distributed poles eastward through Scotts Bluffs to where the construction party had set poles and strung the wire. We rolled into camp at a late hour near Lurjon's Ranch weary and dusty. I drove the cook wagon. On our way to this camp we met at Scotts Bluff the construction gang of men who at this particular time were engaged in digging the pole holes. They were all in excellent health and flow of spirits. I had an enjoyable time with them not having seen them since Creighton and I camped with the Hazards on Lodge Pole creek.

July 30, 1861

In the grey of early morning I was called out of my slumbers by someone at the tent door asking, "Is Brown in"? I gave him to understand that without doubt I was the man he was seeking. He then informed me that his name was Tate and that he had three wagons, thirty-five head of cattle, having lost one, and seven men and that he was engaged in Denver to work for the Telegraph Company by John A. Creighton, and was now ready to commence business. He brought a letter of introduction to me from my brother J. J. From this I learned that he and Tate ~~were~~ had been very well acquainted for about one year. Mr. Tate, whom we all called, Bob Tate, was a genial, whole-souled, companionable man and a gentleman of intelligence. He invited me to take breakfast with him, which I cheerfully accepted and went with him to his wagons. Bob brought out his best and we had a royal good fete. I went with Tate and his small outfit and into the Bluffs to help him get out poles. We stopped about two miles from the cañon and waited for the Hazards to come up which they did about noon time. We gave up the afternoon to cutting poles. About 5 o'clock p.m. quite a severe thunderstorm accompanied with high winds swept over us.

The wood to be found in this region, west and south west of Court House rock, is mostly pitch pine. There is some cedar most of which is small and scrubby. The box elder is fairly well represented. The cotton wood is scarce, but most of the trees are very large.

The soil is composed of sand and clay and as usual is very dry and hard. There are a succession of hills running from south west to north east and are composed of clayey rock. Evidently there once existed a continuous range of hills, but during the ages the action of the elements has leveled it, except the few commanding bluffs remaining, such as Scotts, Court house rock, and Chimney rock. These are left to proclaim what evidently was. They are yet fully three hundred feet in height.[105]

The soil when it has the benefit of water seems productive enough, but the insufficiency of the rain fall excludes this region from agricultural pursuits. This dryness must be overcome in some way before the soil can be tilled. There are a large number of small streams coming out of the cañons but the water soon sinks after leaving the hills. The water from the springs in the cañons is clear and good notwithstanding it holds in solution earthy substances, such as lime and sulphur & c. Many of the streams are highly impregnated with this last substance.

Here the sage brush and grease wood flourish. Several varieties of the cactus exist in large beds and otherwise are promiscuously scattered over the ground compelling man to take heed where he steps. The country is fairly well covered with the various kinds of grasses. These lands on Pumpkin Seed creek and Lawrence Fork and stretching from the Platte River westward around Courthouse rock and still further on into the hills are good pasturage ranges for horses, cattle, and sheep. These last unless well guarded will be an easy and tasteful prey to the wolves.

Went "home" with Bob Tate and camped with him during the night.

This afternoon and evening a rather amusing incident took place. Tate had brought down with him from Denver as an employee a man by the name of "Dave." His full name I never knew. He soon became known only as "Sappi." He was a Missourian and as lazy a fellow as ever matured east of the Mississippi river. The time between each ax stroke was so long that Bob insisted the ax rusted. Bob and I went down to where he was chopping and Tate told him that he looked sick. He replied that he was not feeling very "pert." Bob told him that he had better work on unless he became so sick that he could not well chop any more and that in that event he had better take his ax and go up to the camp. We left him and after passing out of hearing distance from him, Bob says that fellow is not unwell but is lazy and he will go to camp in a short time and sure enough he did. After our day's work was over we went to camp and Bob asked Sappi how he felt. He said he "felt very bad and was right smart sick." Bob asked him if he did not think that some medicine would do him good. Supposing there was none in camp he replied he thought some would do him "a power of good." Bob felt of his pulse and like a long agone [*sic*] country doctor he kept his eye on an old silver watch while timing his pulse. He also intently inspected his tongue and felt of his forehead and in a short time he told Sappi that he had a slight attack of the mountain fever which was liable to prove fatal unless relief was speedily obtained and that he always carried a sure panacea for this dreaded disease. He went to his valise and took out four compound cathartic pills—a very drastic dose—and getting a dipper of water he gave them to his patient and told him to swallow them which apparently he did remarking, "They went down mighty slick." The camp was very quiet that night.[106]

In the morning Billy the cook, killed a mountain sheep and we had some cooked for dinner. The meat of this animal is juicy and sweet and differs very little in taste from deer or antelope flesh. As this was the first fresh meat we had had for a long time the boys pitched in and did ample justice to the repast.[107]

This mountain sheep is considerable larger and more stocky in build than the deer or antelope to which it has a marked resemblance. Their hair or wool is closely alike. They are wonderful climbers of the rocks. Their horns are simply monstrous in size and are shaped very like the horns of an old Merino buck with however only one twist or curve to them. For what purpose they grow so large I cannot tell.

July 31, 1861

Got up at light and anxious to learn the condition of Sappi's health. He had not improved. Bob, distrustful of the efficacy of his medicine, examined the ground where Dave stood when the medicine was administered, and found the pills tramped into the dust. Bob told Sappi that now it was no wonder to him that he was no better, for he had picked up the pills from the earth, that he believed all that ailed him was laziness, but he would give him the benefit of believing him sick and he must take the pills. He did and what was more he retained them. We left Dave to guard camp. The medicine did him a power of good. He grew into an excellent and able bodied man, equal in all respects to any we had. In a short time, forgetting his prayers, he showed a great capacity for swearing, stood up for his rights, threatened to whip one or two, and soon was not the but[t] of his associates. He went back to Denver.

Finished cutting and loading poles at noon. Drew them down to the line and camped about one mile above Lurjon's. Staid over night at Tate's hotel.

August 1, 1861

Commenced the day's labor about light and had our few wagons under motion before sunrise and passed and camped west of Scotts Bluffs about 10 o'clock. Went with Tate and fished in the Platte under the high bluffs. The net result of about an hour's picatorial [*sic*] amusement was one small white perch which I did not catch.

Hazard's loaded teams came up to us about noon. Ed Creighton, who in these days seemed ubiquitous, also came into our camp about noon. He had been down to Julesburgh to attend to hurrying forward to Hibbard supplies and material directly connected with the construction of the line. The day was excessively hot and we took a long nooning. When the trains rolled out Ed, with the buggy team, started for Horse Creek Station—Reynolds' [*sic*] old ranch. We went into camp late.

August 2

Did not get an early start, both men and cattle seemed jaded from the long drive yesterday of twenty-one miles with wagons heavily laden. One of Hazard's oxen died last night. At noon camped at Horse Creek.

On the way up last night set a pole in place of one which had been destroyed by lightning. In the ~~afternoon~~ evening went with Creighton to Hibbard's camp, who has charge of the line builders. A blazing hot day this has been.

August 3, 1861

Last night was ~~a~~ very clear, starlight and beautiful. About 10 o'clock it was cold enough to drive away the sweet singing night birds—the mosquitoes. Slept in a tent with the Frenchmen-Canadians. I sleep decidedly better in the open air. George H. Guy with his teams came up late last night and went into camp below us. Hibbard's men went out to their work bright and early this morning. Took notes from Creighton on matters I was to write for him to eastern parties. Staid in camp until the middle of the afternoon writing letters. After I had completed these went down and fished in the Platte with all the skill I possessed but did not even get a bite. Mr. Selden came up with his two or three teams. Creighton when at Julesburgh employed him. This has been a very hot day. The heat with undulating waves hung over the valley.

August 4, 1861

Mr. Creighton and myself, with our special conveyance—the mules and concord buggy—made a late start for Hazard's train, which had gone up the road loaded with poles. As soon as I reached the train, which was distributing its poles, I immediately began pacing the ground and marking the places to unload a pole. The distance apart the poles were to be placed depended on the character of the ground. If the line was straight and the ground comparatively level, twenty-two to twenty-five poles of large size were used per mile. If the line was full of curves and angles or the surface broken badly, hills and vales, more poles were required. Of course the man who measured the ground had to take all these matters into consideration. The whole line would average twenty-three or four poles to the mile.

We camped near an Indian grave at night. I went up to it and judged from appearances that the dead brave was of some distinction among his tribe—probably Cheyennes. He was not buried, but was laid away to rest on a rudely constructed scaffold. Four poles were set in the ground in form of a square, each about six feet apart. At about the height of seven or eight feet other poles were firmly bound or lashed with raw hide strings, thus fastening the top of the poles securely together. Then on top of these binders or plates other small poles were laid, with a space of three to five inches between them. The body of the dead was then rolled up in a buffalo robe, which was sewed and tied up and made his winding sheet. The body thus prepared for "the happy hunting ground" of the future was placed on the top of this scaffolding with bows, arrows, scalping knives, and tomahawk was left on this aerial lookout to shrivel or mummify in the dry atmosphere. On the ground within the four upright poles two ponies had been killed and from appearances several dogs. These also formed a part of his post mortem equipment in his spiritual existence of savagery.

On this trip I also saw Indian burials in which trees had been used instead of poles. I do not think I saw an instance of interment of the Indian dead. This method of disposing of their dead is one which would naturally be adopted by the savages of the plains and hills of the west. They had no implements for digging graves and in that dry hard soil if they had proper tools, the graves would be likely to be shallow and the body easily dug up by the wolves which, in all probability, are very fond of dead Indian. Placed on these poles it is secure from the jaws of these ravenous animals.

Late at night Creighton came to Hazard's camp and staid all night. He and I slept on the ground together by the side of the wagon. It was a hot day and the dancing heat made one of those wonderful mirages occasionally seen in the valleys on the plains.

August 5, 1861

Creighton remained during the day with Hazard's and Hibbard's trains. Commenced my day's labor early in the morning, which was marking for the distribution of poles by Hazard. Went into camp at noon near Beauvais ranch. Both Hibbard and Hazard's trains were resting together. Creighton connected a transmitter to the wires we had strung and sent messages east, of a friendly and business nature.

Went down with the men to Beauvais ~~with~~ to do some trading for them. Made for the several men, who absolutely needed what we bought, a bill of over $125.00. The price of everything for sale was so high that we bought just as [few] goods as possible with regard to the comfort of the men. We purchased nothing outside of wearing apparel and tobacco. Our cattle strayed while grazing on grounds claimed as a part of the fort's Reservation and was notified to remove them.[108] We cheerfully complied with the notification. Case came up and went into camp with us at noon.

Quite an amusing incident took place at this encampment. Along side of the overland emigrant road and a short distance from Beauvais there was a fairly good sized band of Cheyenne Indians in camp. After I returned from trading, Creighton proposed that a few of us pay a visit to our neighbors. The operator who traveled with Hibbard was taken with us. He carried with him a battery and other requisites for administering electrical shocks. When we reached the camp we found "a squaw man" who could serve as interpreter between ourselves and ~~visitors~~ the Indians. Creighton and Hibbard told them what we were employed at. That lightning ran along the wire and talked and made them in a hazy way understand how dangerous it would be to interfere with our work. Creighton, then, practically, to illustrate what he had been saying, proposed to give them a shock from the battery. This being arranged Hibbard generated a very powerful electrical charge, joining hands with the Indians, squaw and buck, Hibbard "let her go" and there was pranc-

Henry Farny, *The Song of the Talking Wire,* 1904, oil on canvas. Bequest of Charles Phelps and Anna Sinton Taft, Taft Museum of Art, Cincinnati, Ohio. Photographed by Tony Walsh, Cincinnati, Ohio

ing then and there among the Indians. They did some talking and looked upon Mr. Creighton as "a big medicine man."[109]

When Hibbard commenced stringing wire that afternoon it so happened that he had to run it directly over the tepes of this camp of Indians. In hot haste the squaws <u>and also the bucks</u> tore down their lodges and decamped.

Where we quit work that night our telegraph line was complete to three miles west of Fort Laramie. Hazard with eight teams started this afternoon for Julesburgh to bring forward wire, insulators, & c, while the rest of his train returned to Horse Creek to bring forward poles.

Between three and four p.m. there was a short but severe thunderstorm attended with a heavy wind.

After supper started with Creighton and an employee, called California, a forty-niner on his way to his home in New York, for Center Star.[110] We went in the Concord buggy. The night was very dark and at times we could not distinctly see our mules. With difficulty we kept in the road. The mules however took us through safely. We kept a steady rein on them and let them have their way. We arrived at Center Star ranch about midnight. Learned that George Guy was in camp a mile or so to the eastward. When the "overland stage" came up I boarded the same for Horse Shoe Creek Station which I reached about 6:30 o'clock the next morning.

August 6, 1861

Had an extra good breakfast this morning at the station. Upon inquiry ascertained that Tate had not reached this point. Tried to make arrangements to store the wire, insulators, & c with which Tate was laden at this station but the man in charge refused to receive them. Whether he was indisposed to accommodate from his own inclination [or] was in mortal terror of Slade, who was division superintendent of the Overland Company, I did not know.[111] Went on foot down to where Jim Creighton was in camp cutting poles. He had already cut and distributed from this point—the foot hills of Laramie Peak—1,075 number one poles and then had cut and shaved, ready for hauling, about 450 more. Took dinner with Jim and then walked back to the station. I found a few very nice specimens of mica and gypsum on my way back.

When I reached the station I was hungry and very nearly tired out from the labors of the past few days.

August 7th, 1861

Took a late breakfast and lay around the station waiting for Bob Tate's train to come up. Becoming impatient over the tardiness of his arrival I started on foot down the road to meet and get his train into a livelier motion. Met him about one mile east of the station. On returning to the ranch I put aboard the train my baggage and then pushed on as rapidly as possible for the foot hills of Laramie Peak in the neighborhood of Jim Creighton's camp. Someone stole my ax which I had with my baggage at the station. This I never recovered. Reached the pinery about 2 o'clock p.m. and went into camp. After dinner went to getting out poles.

August 8th, 1861

Worked all day in the woods cutting, peeling, and piling poles. There was a thunderstorm late in the afternoon.

When Creighton and I were last at the camps around Scotts Bluffs, he united in one train the two small ones respectively under the supervision of Joe Creighton and Bob Tate and placed the new train in the joint charge of Joe and Bob. The union of the two small trains was the thing to do, but it was short sighted policy to place it under the joint control and command of these two men. Ed knew this, but he could not find it in his kind heart to do that which would seemingly cast any reflection on the competency of his brother Joe. Bob was hired by John A. Creighton to have charge of a train, and thus Ed could not reduce him to the ranks of an ordinary hand. Bob was competent to take charge of a train where his duties as wagon master or captain were the same day in and day out, as in freighting from the Missouri River to Denver, but he was not clearly a success in our enterprise when the chief man of the train had to be at all times an active committee on ways and means. To sum up Joe Creighton, his character and his abilities to accomplish anything in a few words, I must say he was absolutely and largely unqualified to give any results whatever. He was lazy mentally and physically and so destitute of energy that he neither did nor could do anything. He was a human sloth. Yet everybody liked Joe; he was such a good hearted fellow.

As could well be expected, this new train, under a dual headed management, was a marked and re-marked failure. Joe could do nothing and Bob felt that to assume control and virtually exclude Joe would be an act ungracious to Mr. Creighton and not justified and consequently this train did little, under this double generalship, unless to execute the commands of Creighton given by him or through his authorized agents. In a short time both man and beast became demoralized and how to do nothing became the study of all connected with it.

Later on it became my fortune or misfortune to assume control of this train at Rock Creek east of the famous South Pass and to remain in charge of it until we arrived at Green River. Joe was along, it was Joe Creighton's train, but I assumed the command. Some few of the men were inclined to be ugly at my assumption and exercise of authority, but in a short time they fell into the new order of running the train and all worked harmoniously and energetically. No one could ask for better men and they cheerfully imbibed the spirit of trying to do a little better than the other trains.

August 9, 1861

Worked in getting out poles til 9 a.m. and then went with Jim Creighton to assist him in distributing his loads up the road. We did not return to camp until after nine o'clock at night. I found the Tate train had decamped. Ed had been there and ordered them down to the station to unload and return the next morning. Slept on the ground in the open air. How clear the night! How brightly the stars do twinkle and glisten in the cloudless azure sky! [The diary ends without further explanation.]

First Telegraph Line across the Continent: Charles Brown's 1861 Diary

THE FIRST TELEGRAPHIC MESSAGE FROM CALIFORNIA.

Illustration from *Harper's Weekly,* November 23, 1861. Library of Congress Prints and Photographs Division

Epilogue

Paradoxically, as the telegraph tied the continental United States together east to west, the secession crisis of 1860-61 severed American unity and brought on the Civil War between the North and South. The telegraph helped transform that conflict into the first modern war. As Maj. Gen. A. W. Greely proclaimed, "The exigencies and experiences of the Civil War demonstrated, among other theorems, the vast utility and indispensable importance of the electric telegraph both as an administrative agent and as a tactical factor in military operations."[112] Similarly, Maj. Gen. William Tecumseh Sherman asserted that "the value of the magnetic telegraph in war cannot be exaggerated." For the Union, the telegraph office in the War Department in Washington, D.C., became "the nerve center of the war."[113]

Prior to the Civil War, the U.S. Army did not maintain a separate signal corps; thus, it had few enlisted men with operational knowledge of the telegraph. To obtain the needed operators, it established the U.S. Military Telegraph Corps. A quasi-military organization, the approximately 1,200 operators (including seventeen-year-old Thomas Edison in the office at Memphis, Tennessee) remained civilians. The telegraphers suffered nearly the same casualty rate as soldiers, with about one out of twelve being captured, killed, or wounded.[114]

A lack of equipment and operators handicapped the Confederacy throughout the contest. While the military constructed more than fifteen thousand miles of wire for the Union, the South managed to add only about one thousand miles. Therefore, the Confederacy maintained an undersized field telegraph service and depended heavily on existing commercial lines. Each of the belligerents coded their messages; Union cryptographers deciphered the Confederate symbols, but the South failed to gain such an advantage. Scouts armed with a small roll of wire and a "pocket" instrument sometimes infiltrated deep into enemy territory to patch into telegraph lines to obtain vital information on battle plans and to communicate deceptive information to confound rival commanders.[115]

Through the telegraph system, the U.S. War Department controlled the long-distance transportation of men and supplies and coordinated the plans and activities of individual units, armies, and commands. President Abraham Lincoln did not install a telegraph office in the White House; according to Major-General Greely, the commander-in-chief preferred to visit the office at the War Department across the street, spending "long hours" there and using it as a "retreat both for rest and also for important work requiring undisturbed thought and undivided attention."[116] Historian Doris Kearns Goodwin emphasized that "On the nights he did not spend with [Secretary of State William Henry] Seward, Lincoln found welcome diversion in the telegraph office, where he could stretch his legs, rest his feet on the table, and enjoy the company of the young telegraph operators." One of those young operators in 1862-63 was Edward Rosewater, later the founder and proprietor of the *Omaha Daily Bee.*[117]

Unlike the telegraph lines built and used by the armies for tactical and strategic purposes, the transcontinental line gave the Union important political and economic advantages that helped it win the war. The transcontinental telegraph enabled the Lincoln administration to maintain direct communication with the loyal states and territories of the West, most of which also produced the gold and silver that was critical to financing the war. Protecting the transcontinental telegraph line was deemed so important that soldiers were stationed along its route throughout the war, despite the heavy demands for Union troops on the major battlefronts east of the Missouri River.[118]

The completion of the transcontinental telegraph in 1861 had signaled the demise of the iconic, but short-lived (sixteen-months) Pony Express. Cyrus Field laid the first trans-Atlantic telegraph cable in 1858, but it soon failed. In 1866 Field succeeded in laying a new Atlantic cable and by 1900, thirteen undersea telegraph cables traversed the Atlantic.[119]

Internal development in the United States manifested similar rapid expansion. After the Civil War, the newly organized Army Signal Corps became responsible for building and maintaining a military telegraph system. By 1881 more than five thousand miles of wire tied together the nation's far flung military posts and headquarters. A primary assignment of the military telegraphers was reporting weather conditions to the War Department three times a day. By 1880 the army had established 110 weather stations.[120]

In the field, the troops continued to use signal flags and torches and in 1888, the army adopted the heliograph, a device that reflected the sun. The geographically isolated, small-scale warfare with the American Indians prohibited the use of Civil War-style field telegraphy. As settlement filled the interior West, telegraph lines followed. The army increasingly had commercial telegraph lines at its disposal, but it limited their use to emergencies because of the expense: a telegram included a basic charge for the first ten words, plus an added per-word charge thereafter, and the total cost depended upon the distance of the transmission (e.g., the early transcontinental transmissions, San Francisco to St. Louis, cost $5 for ten words, then 45 ¢ for each word thereafter).[121]

Such an emergency arose in the early morning of July 5, 1876, when John M. Carnahan, telegrapher at Bismarck, North Dakota, site of the Seventh U.S. Cavalry's headquarters post of Fort Abraham Lincoln, was awakened from his bed to send the news of Custer's defeat at the Little Bighorn. Carnahan remained at his station for twenty-two hours, transmitting news of the battle as it trickled in. Similarly, at the Wounded Knee Massacre on South Dakota's Pine Ridge Reservation in December 1890, "the click of the [telegraph] operator's instrument almost mingled with the rattle of the rifle shots."[122]

Two decades earlier, the telegraph had alerted Americans to another event with dramatic consequences for the nation's future. On May 10, 1869, the *New York Times* reported, "At 2:20 this afternoon, Washington [D.C.] time, all telegraph offices in the country were notified by the Omaha telegraph office to be ready to receive the signals corresponding to the blows of the hammer that drove the last spike in the last rail that united New York and San Francisco with a band of iron." Telegraphers at Promontory Summit, Utah, connected wires to the sledgehammer and to the golden spike; thus, each strike became "a tick on the nationwide telegraph network." Telegraph lines routinely paralleled railroad tracks, and railroad and telegraph stations usually conjoined, especially in small towns.[123] By the end of the nineteenth century, the telegraph system looked like a spider web draped over the country. As Tom Standage's book title suggests, it became *The Victorian Internet,* its dots and dashes providing many of the services contemporary computer users access through the binary system of zeros and ones.

The telegraph system worked in a similar fashion as those established by the post office, railroads, or contemporary airline hubs. Messages moved directly between large cities, where relay stations received and retransmitted them to specific destinations. In 1872 Joseph Stearns increased the flow of communications with the invention of the duplex, which allowed two messages to stream over the same wire simultaneously. Three years later Thomas Edison doubled that potential with the

THE PROGRESS OF THE CENTURY.

THE LIGHTNING STEAM PRESS. THE ELECTRIC TELEGRAPH. THE LOCOMOTIVE. THE STEAMBOAT.

Published by Currier & Ives, 1876. Library of Congress Prints and Photographs Division

quadruplex, permitting four simultaneous transmissions. By 1900 the Chicago relay office employed 880 operators per shift that handled more than two million messages monthly. Between 1866 and 1900 the number of annual communiqués sent over Western Union lines soared from 5.8 million to 63.2 million. The cost per telegram dropped to an average of 30 cents.[124]

The railroads used the telegraph to dispatch and control trains until the 1950s. Moreover, in 1865 the U.S. Naval Observatory in Washington, D.C., began supplying time signals to Western Union. The railroads immediately began to use the information for scheduling. In 1883 the U.S. government established standard time zones to systematize the process. Similarly, the telegraph permitted the reporting of distant events within hours of their occurrence. As early as 1848 the Associated Press (AP) had organized to send and receive dispatches over commercial lines. After 1875 it began to establish its own network of wires connecting member newspapers. In the 1880s Walter P. Phillips of the AP invented a code that eliminated spelling out words and phrases commonly used by reporters (e.g., "c" = see; "d" = in the; "7" = that is, presaging the contemporary jargon of e-mail and texting). In 1907 E. W. Scripps organized the United Press (UP) as a competitor, and in 1909 William Randolph Hearst the International News Service (merged with UP in 1959 to form United Press International).[125]

The telegraph industry experienced the evolution of similar competition with the founding of the Postal Telegraph Company (not associated with the U.S. Post Office) in 1881 and the American Telephone and Telegraph Company (AT&T) in 1885. Furthermore, the coordinating function of the telegraph abetted the growth of large businesses nationwide, as well as spawning novel enterprises. In 1867 the New York Stock Exchange inaugurated the stock ticker service. Five years later Western Union instituted its money transfer system using coded messages between offices. Both of these services unleashed a multitude of criminal activities involving the bribing of operators, tapping of lines to gain information, breaking business codes, and message scams. More appropriately, in 1910 a group of retail florists seeking to ensure "customer satisfaction on out-of-town flower deliveries" founded the Florist's Telegraph Delivery" (FTD).[126]

Moreover, flirtations and romances blossomed over the wires because of the presence of female operators. Women entered the occupation in the 1840s, and by 1870 they comprised one-third of the employees in the main telegraph office in New York City. Not all of the connections remained honorable, leading the journal *Electrical World* in 1886 to publish a cautionary tale entitled "The Dangers of Wired Love."[127]

Similarly, the owners of major league baseball franchises feared negative effects from the infringement of modern communications. Thus, they banned radio broadcasts of games, apprehensive that they would lessen attendance at the ballparks. In 1934 most owners partially relented, allowing telegraphed descriptions of the contests to be sent to outlying towns for broadcast on the radio. Thus, future president Ronald Reagan had a brief career in Des Moines, Iowa, doing the play-by-play of Chicago Cubs games over the radio. Telegraphers developed a shorthand to allow rapid transmissions to keep up with the on-the-field action (e.g., SIC = strike one called; NTG AX = no runs, no hits, no errors, none left).[128]

Reagan's broadcasts benefited from the recent introduction of the teletypewriter and teleprinter. Numerous inventors contributed to the development of the mechanisms, which made their debut in the mid-1920s. A circuit connected machines so what was typed on one was retyped at the other end. The message came out on pre-glued strips that were then pasted on the telegram sheet.[129] While the teletypewriter doomed Morse code, the telegraph remained the favored form of long-distance communication until the 1950s because of the limitations of the telephone network.[130]

As early as the 1870s Alexander Graham Bell had worked on developing the "harmonic" telegraph that would send multiple sound pitches simultaneously, thus increasing efficiency and volume. His ultimate invention, the telephone, at first garnered the moniker "speaking telegraph." The telephone system did not acquire its first transcontinental line until 1915, and the service remained expensive and often of poor sound quality. Although Guglielmo Marconi patented a wireless signaling device in 1896 and incorporated the Marconi Wireless Telegraph Company in London the following year, it would take decades to develop the ever-more-powerful transmitters and antennae necessary to compete with the wired telegraph.[131]

AT&T and the Bell system created a telephone teletypewriter equivalent (telex), and experimented with radio relay systems, but the Great Depression and World War II slowed the advance of long-distance telephone service. During that era, the Postal Telegraph Company earned about 15 percent of the business, while AT&T garnered 18 percent, and Western Union reaped a 64 percent share. In 1945 Western Union reached its peak of 236 million messages, but then its traffic began a steady decline.[132]

Beginning in the 1950s the new technologies expanded exponentially, so that by 1970 Western Union delivered only 69 million telegrams. After decades of mounting losses, mergers, and buyouts, the Western Union Telegraph Company reorganized in 1988-89 as the Western Union Company to specialize in the profitable business of money transfers. Subsequently, on January 27, 2006, Western Union discontinued its telegram and commercial messaging services. Some small companies continue to provide the service, but telegraphy has dwindled to an infinitesimal part of the communication business now dominated by computers, cell phones, texting, and tweeting, *ad infinitum.*[133]

From its 1837 patent by Samuel F. B. Morse through its demise more than a century later, the electric telegraph evolved to become the world's primary means of long distance communication. Like modern "smart" phones and other internet-compatible devices, telegraphy was the communications revolution of its day. When the first telegraph line reached Brownville, Nebraska Territory, in 1860, the local newspaper exultantly proclaimed what could as well be the watchword of the digital age: "By Lightning, Telegraph to Brownville, Time and Space Annihilated."[134]

Notes

[1] Brown's handwritten manuscript is among the records of the Western Union Telegraph Company, Archives Center, National Museum of American History, Smithsonian Institution, Washington, D.C., and is published by permission of the Smithsonian Institution.

[2] Lewis Coe, *The Telegraph: A History of Morse's Invention and Its Predecessors in the United States* (Jefferson, N. Car., and London: McFarland and Co., Inc., 1993), 1-3.

[3] Ibid., 6; Tom Standage, *The Victorian Internet* (New York: Walker and Co., 1998), 8-14.

[4] Coe, *The Telegraph,* 15.

[5] Ibid., 15, 29; Standage, *The Victorian Internet,* 26-27.

[6] Thomas Streissguth, *Communications: Sending the Message* (Minneapolis: The Oliver Press, Inc., 1997), 31-32; Coe, *The Telegraph,* 67; Standage, *The Victorian Internet,* 28-29.

[7] Coe, *The Telegraph,* 16, 31, 67; Standage, *The Victorian Internet,* 37-39; Streissguth, Communications, 33.

[8] Coe, *The Telegraph,* 32; Standage, *The Victorian Internet,* 42-48, 53-65.

[9] Dennis N. Mihelich, *The History of Creighton University,* 1878-2003 (Omaha: Creighton University Press, 2006), 3-8. James W. Savage and John T. Bell, *History of the City of Omaha, Nebraska* (New York: Munsell and Co., 1894), 112, credited Creighton with building the telegraph from St. Louis to Omaha in 1860.

[10] Robert Luther Thompson, *Wiring a Continent* (Princeton, N. J.: Princeton University Press, 1947), 354 and Appendix 15, 515-17.

[11] Thompson, *Wiring a Continent,* 359-63; Brown Diary, 1.

[12] Mihelich, *History of Creighton University,* 8.

[13] Brown Diary, 7.

[14] Coe, *The Telegraph,* 39-41; http://www.ieeeghn.org/index.php/transcontinental_telegraph_line_%28US%29.

[15] Thompson, *Wiring a Continent,* 363-68.

[16] Mihelich, *History of Creighton University,* 8-12.

[17] Arthur C. Wakeley, Omaha: *The Gate City and Douglas County Nebraska* (Chicago: The S. J. Clarke Publishing Co., 1917), 12-15; Evening *Omaha World-Herald,* Apr. 26, 1897.

[18] J. Sterling Morton, succeeded by Albert Watkins, *Illustrated History of Nebraska* (Lincoln: Jacob North and Co., 1905), 1: 510.

[19] Ibid.; *Omaha Bee,* June 13, 1886 [the article spelled the brother's first name "Lewis," as did a reference to him in Savage and Bell, *History of Omaha,*116, but Margaret's marriage certificate spelled her father's name "Louis"]. The cover sheet to the diary identified Margaret as the niece of Charles H. Brown. U.S. Manuscript Census, Population Schedule, 1900, Douglas County, Nebraska.

[20] Evening *Omaha World-Herald,* Apr. 26, 1897.

[21] The line from St. Joseph reached Brownville in late August 1860, Omaha in early September, and Fort Kearny by late October. *Nebraska Advertiser* (Brownville), Aug. 30, 1860; *Huntsman's Echo* (Wood River Center), Sept. 13, 1860; *Council Bluffs* (Ia.) *Nonpareil,* Nov. 10, 1860.

[22] Aspinwall is the modern city of Colon.

[23] For example, the Omaha *Daily Telegraph* reported on May 26 that the steamboat *West Wind* had delivered 22,870 pounds of wire for the Pacific Telegraph Company.

[24] Lillian Springs was a ranche about forty-four miles west of Julesburg on the South Platte route to Denver, according to a table of distances and list of road ranches published in the *Daily Telegraph* of Apr. 9, 1861, and subsequent issues. Evidently Creighton was acquiring freighting teams wherever he could find them. In May 1861 Creighton bought out the Council Bluffs firm of Baldwin and Dodge, acquiring ninety yoke of oxen and fifteen wagons for $8,000 cash. *Daily Telegraph,* June 1, 1861.

[25] Brown inserted the date December 28, 1890, in brackets within this sentence, leading the editors to conclude that it signals when he began writing his introduction and transcribing his field diary as herein.

[26] Cottonwood Springs, southeast of present North Platte, was at the intersection of a north-south Indian trail and the Platte Valley trail to the west. A spring, and cedar timber in the nearby canyons, made it a popular point for road ranches and later for a military post known as Fort Cottonwood, later renamed Fort McPherson.

The *Daily Telegraph* of June 2, 1861, reported that "the first of Mr. Creighton's trains left yesterday, twenty wagons, with an average weight of 5,600 pounds of telegraphic freight." This belies Brown's comment in the next paragraph that James Creighton's train, "the second to move out," left Omaha about May 10. According to the *Daily Telegraph* of June 6, train number two, consisting of eighteen wagons, left town on June 5. It is possible, of course, that the newspaper simply did not notice freight outfits leaving earlier with telegraph supplies for the construction to begin at Julesburg, given the large number of trains departing in the regular trade to the Colorado gold fields. May departures may have been supplies to complete the line from Fort Kearny to Julesburg. The *Daily Telegraph* of May 29, 1861, reported that "Mr. Porter" (probably Omaha commission merchant Henry) had started twelve wagons from Omaha to Cottonwood Springs and would return to take out the rest of the supplies to complete that section of the line. The same issue noted that Creighton was waiting for eighty wagons to be loaded with supplies for the line from Julesburg to Salt Lake City. About 160 tons of supplies had been received in Omaha.

[27] Rawhide Creek is a tributary of the Elkhorn River located east of present Fremont. The stream was named for the skin boats that Indians and fur traders made to cross it when the water was high. About 1850 a story gained currency in the West and elsewhere that the stream was where an overland emigrant murdered an Indian woman without provocation and her tribe then skinned the murderer alive as retribution. Although numerous versions of the tale, naming numerous victims, have surfaced and then been told and retold, the story is a myth. See James E. Potter, "The Legend of Rawhide Revisited," *Nebraska History* 85 (Fall 2004): 128-39.

[28] Brown says this man's name was Wells in his entry dated July 17, 1861. Incidents in which overland travelers shot themselves by careless handling of firearms are numerous and they span the entire period from 1841 through the 1860s. Many of the casualties were civilians from farms or cities, who did not have much previous experience with guns. Other victims were more experienced "frontiersmen," who had likely developed a cavalier attitude toward a tool that most of them handled daily. The incident Brown describes is a classic example of the type of accident that was all too common. For a review of firearms accidents suffered by overland emigrants see James E. Potter, "Firearms on the Overland Trails," *Overland Journal* 9 (1991): 2-12

[29] The foregoing section appears to be Brown's introduction, while the section after this break seems largely transcribed from his daily diary. Apparently Brown had taken the stagecoach from Omaha to catch up to the trains hauling the telegraph supplies.

[30] Adolphus M. Hart's novelette was published ca. 1854. See Lee Ann Sandweiss, ed., *Seeking St. Louis: Voices from a River City, 1670-2000* (St. Louis: Missouri Historical Society Press, 2000), 173.

[31] Esquire "Squire" Lamb and his wife, Caroline, operated a road ranche and Western Stage Company station located east of the present town of Wood River. The Lambs came to Nebraska Territory from New York State in 1858. Letters by Lamb and other family members are found as MS1561 at the Nebraska State Historical Society, Lincoln.

[32] Johnson had previously published newspapers in Council Bluffs, Iowa, and in Omaha. His story is told in Benjamin Pfeiffer, "The Role of Joseph E. Johnson and His Pioneer Newspapers in the Development of Territorial Nebraska," *Nebraska History* 40 (June 1959): 119-36.

[33] "Adobe Town," better known as "Dobytown" and officially as Kearney City, was the notorious civilian settlement just beyond the Fort Kearny military reservation to the west. It is not to be confused with modern Kearney north of the Platte River.

[34] Brown refers to a Latin phrase translated as, "Let him not go out of the kingdom." In legal terms it means a restraining order to prevent a person from leaving a particular locality or the jurisdiction of a court without posting bond. James M. Woolworth was a prominent Omaha attorney. See Morton-Watkins *Illustrated History of Nebraska,* 1:775. Ben Holladay (1819-87) acquired the Central Overland California and Pikes Peak Express stagecoach line from Russell, Majors, and Waddell in 1862 and under his direction, stagecoach service from the Missouri River to the West Coast reached its zenith. Holladay sold the company to Wells Fargo in 1866. Dan Thrapp, *Encyclopedia of Frontier Biography,* 3 vol. (Lincoln: University of Nebraska Press Bison Books, 1991) 3:669.

[35] Lt. Daniel Woodbury of the army engineers selected the site in the fall of 1847, but the first troops did not arrive to begin building the post until spring of the next year.

[36] No source has been found to indicate what may have been wrong with Brown's hand.

[37] This, of course, was soon after the outbreak of the Civil War in April 1861. See Governor Samuel Black to Secretary of War Simon Cameron, April 27, 1861, in *The War of the Rebellion: A Compilation of the Official Records of the Union and Confederate Armies,* 128 vol. (Washington, D.C.: GPO, 1880-1901), Ser. 3, v. 1, 123.

[38] The Western Stage Company had begun weekly mail service from Omaha to Denver and back via Fort Kearny in September 1860. *Huntsman's Echo,* Sept. 13, 1860. The Omaha *Daily Telegraph* reported the beginning of daily mail service from Omaha to Denver as of Aug. 5, 1861. When Brown transcribed his field diary, he inserted the recollection of his 1862 attempt to serve legal papers on Ben Holladay. When Brown wrote his original entries in June 1861, Holladay had not yet taken over the Central Overland California and Pikes Peak Express Company. See William E. Lass, *From the Missouri to the Great Salt Lake: An Account of Overland Freighting* (Lincoln: Nebraska State Historical Society, 1972), 122-23.

[39] George R. "Doc" Smith, an Ohio native, moved to the Omaha area about 1856. He served on the Omaha City Council, 1871-72, and as Douglas County Surveyor, 1873-75. He was noted for his frequent letters to the newspapers about public affairs. Smith's letters and other writings emanated from a small study or "den" he constructed adjacent to his house that bore the sign, "Dox Box" above the door. A. T. Andreas, *History of Nebraska* (Chicago: Western Publishing Co., 1882), 692, 695, 707; Morning *Omaha World-Herald,* Jan. 11, 1901.

[40] Talbot operated one of about a dozen stores at Kearney City, which had forty or fifty buildings, according to Joseph E. Johnson, who visited there in late October 1860. *Huntsman's Echo,* Nov. 2, 1860.

[41] Before the Homestead Act of 1862 took effect on Jan. 1, 1863, the Pre-Emption Act of 1841 was the principal method for an individual to acquire public land in the West. It was essentially a contract to purchase up to 160 acres at the government price of $1.25 per acre, with the proviso that the land could be "pre-empted" in advance of the survey and the boundaries adjusted later.

[42] One of the firms at Kearney City in the fall of 1860 was McDonald & Young. *Huntsman's Echo,* Nov. 2, 1860.

[43] One surmises that Brown had taken the Western Stage Company coach from Omaha to Fort Kearny. To continue west, passengers then had to transfer to the coaches of the COCPP, which originated at Atchison, Kansas. If the "through" coaches were full when they reached Fort Kearny, passengers might have to wait several days to get a seat. This may explain why Brown sought another way to get to Julesburg. Frank A. Root and William E. Connelley, *The Overland Stage to California* (Topeka: The Authors, 1901), 205-6.

[44] There were several ranches in this area and the Seventeen-mile Point, as its name implies, was about seventeen miles west of Fort Kearny. Nearby was Moses Sydenham's ranche, also known as Hopeville. Sydenham was the long-serving postmaster at Fort Kearny, whose mother and siblings emigrated from England and lived at Hopeville. When the telegraph reached Fort Kearny, the station was first housed in Sydenham's post office. Morton-Watkins *Illustrated History of Nebraska,* 1:98-99.

[45] Daniel S. Parmalee shot and killed Thomas Keeler on Dec. 5, 1874, according to Andreas, *History of Nebraska,* 710, 789, 817. Harrison Johnson's *History of Nebraska* (Omaha: Henry Gibson, 1880), 291-92, dates the event as Dec. 10, 1874. Johnson credits the altercation to a dispute arising from Keeler's cattle having trespassed on Parmalee's land. Although Brown says the ranche west of Fort Kearny was operated by Thomas Keeler's brother, it is listed as "Eight Mile Point by Thomas Keeler" in the guide to "Route, Camping Places, Ranches, etc. etc." published in the *Daily Telegraph,* Apr. 9, 1861. Thomas

Keeler, age twenty-eight, a "trader" born in Ohio, is listed in the 1860 census of Dodge County. Population Schedule, Nebraska Territory, Eighth Census, 1860, National Archives Microfilm Publication M653, roll 665 (hereafter 1860 census).

[46] The exact particulars of this incident are unknown. A letter from the Pawnee Reservation dated Oct. 13, 1860, mentioned that six hundred Sioux had crossed the Platte near Cottonwood Springs on their way to attack the Pawnees. William Stolley of the Grand Island German settlement was on a buffalo hunt near Fort Kearny in the late fall of 1860 and noted that the officers there reported that a thousand Sioux were then hunting on the Republican River some forty miles to the south. Clearly the Sioux were roaming the area around this time and it may have been some of these Indians with whom the whites fought near Plum Creek. Omaha *Daily Nebraskian,* Oct. 20, 1860; Stolley, "History of the First Settlement of Hall County, Nebraska," *Nebraska History* Special Issue (April 1946): 35-36.

[47] The whereabouts of Brown's 1860 account is unknown. In his June 24, 1861, entry he mentions that he had been an ox driver for John A. Creighton in 1860. The ranche here was operated by Thomas "Pat" Mullally and also known as Willow Island. It was located about eight miles southeast of present Cozad. Merrill J. Mattes, *The Great Platte River Road* (Lincoln: Nebraska State Historical Society, 1969), 273.

[48] Mattes, ibid., has the distance from Mullally's ranche to Dan Smith's ranche as about nineteen miles, but he mentions an 1860 traveler who placed "Smithe's Ranche" nine miles from Mullally's and the 1861 route guide in the Omaha *Daily Telegraph* also shows "Smith's Ranche" at about the same distance. While nineteen miles seems like a long distance to travel between a "late start" and "nooning," there is no way to know whether "nooning" actually took place at noon. Trail travelers would be likely to proceed as long as light permitted, or until they reached the destination they had in mind. Trail accounts often vary significantly in the distances cited between ranches.

[49] The Pawnee may have once numbered as many as ten thousand, but by 1861 the population had been much reduced by epidemics of smallpox and cholera beginning in the early nineteenth century. In 1833 the Pawnee ceded their Nebraska lands south of the Platte River and agreed to give up the chase in return for government support and protection. The government, however, failed to meet its obligation to protect the Pawnee, subjecting them to forty years of unrelenting attacks, mainly by the Sioux. See Douglas R. Parks, "Pawnee," in *Handbook of North American Indians,* Vol. 13, Pt. 1: Plains, ed. Raymond J. DeMallie (Washington, D.C.: Smithsonian, 2001), 520-21.

[50] Hair ornaments such as Brown describes were likely made of German silver, an alloy of nickel, zinc, and copper.

[51] This ranche, operated by brothers John and Jeremiah Gilman, was southwest of today's Gothenburg. Its history is found in Musetta Gilman, *Pump on the Prairie: A Chronicle of a Road Ranch,* 1859-1868 (Detroit: Harlo Press, 1975).

[52] Several ranches sprang up here, perhaps the most notable operated by Charles McDonald. In the fall of 1863 nearby Fort Cottonwood was constructed by soldiers of the Seventh Iowa Volunteer Cavalry, later designated Fort McPherson. See Eugene F. Ware, *The Indian War of 1864* (Lincoln: University of Nebraska Press, 1960), who tells of the fort's construction. In the manuscript to this point, Brown periodically inserted dates from June 12, 1891, through June 30, 1891, likely representing when he wrote or transcribed these pages. Here, he inserted the date of January 31, 1894. All these insertions have been deleted.

[53] Sylvanus Dodge, father of Civil War general and Union Pacific civil engineer Grenville Dodge, arrived in Nebraska Territory in 1855 and claimed land on the Elkhorn River northwest of Omaha. In 1859 he established a ranche and store near Cottonwood Springs, later returning to the Omaha area. Morton-Watkins, *Illustrated History of Nebraska,* 1:638-40.

[54] Morrow's Junction Ranche, near the confluence of the North and South Platte rivers, was one of the best known such establishments along the trail. Its proprietor was known far and wide as the "Iron Man of the Plains." Many travelers stopped there and left descriptions of Morrow and his ranche. See Mattes, *Great Platte River Road,* 276-77, for several examples. With the decline of wagon travel after the transcontinental railroad was completed, Morrow moved to Omaha, where he filled contracts to supply military posts and Indian agencies. He died there on July 7, 1876, while still in his early forties. Contrary to Brown's statement that he died poor, one of his obituaries said Morrow "was worth a handsome property" with $10,000 of life insurance. *Omaha Daily Herald,* July 8, 1876; *Omaha Daily Republican,* July 8, 1876.

[55] At this point the bluffs hugged the south bank of the Platte, forcing wagons to leave the river bottom and traverse the high ground for some distance. Wagon ruts from the trail days can still be seen at the Nebraska Department of Roads eastbound I-80 rest area located atop the bluffs.

[56] Dorsey's Ranche was also a Pony Express station. Harrison Johnson came to Omaha in 1854. He is best remembered for writing the first history of Nebraska in 1880. Morton-Watkins, *Illustrated History of Nebraska,* 1:234. Cyrus Morton settled near Fremont in 1856, and was a member of the Platte Valley Claim Club. Andreas, *History of Nebraska,* 633. Many settlers engaged

in small-scale freighting to the Rocky Mountain gold camps during this period.

[57] The *Daily Telegraph,* Apr. 9, 1861, noted that an unnamed Omaha resident was planning to take cats to Denver "in a few weeks." If Plumbach made the profits Brown mentioned, it would not be surprising if he decided to take out another shipment of the rat killers the next year.

[58] Louis "Giraux," occupation trader and a native of Canada, is enumerated in the 1860 census of Nebraska Territory for the "Platte River Settlement." "Dixon" may be Peter (Sam) Dion, who is listed in the 1860 census for Fort Laramie. Ware met Dion near there in 1864. The region encompassed by the "settlement" extended from just east of present Paxton, Nebraska, along the South Platte River into today's Colorado, terminating approximately south of Sidney, Nebraska. Those living there were ranche operators, or station keepers, laborers, tradesmen, and teamsters employed by the Pony Express or the stagecoach company. Because these two men were sons-in-law of trader G. P. Beauvais (see n.59) it is reasonable to assume that the latter had a financial interest in this ranche. Iowa Cavalryman Eugene F. Ware, who spelled the proprietor's name "Jereux," said the ranche had been abandoned by 1864. Ware, *The Indian War of 1864,* 178-79, 255.

[59] Geminien P. Beauvais was born in St. Genevieve, Missouri, in 1815. After working in the Upper Missouri and Rocky Mountain fur trade during the 1830s and 1840s, Beauvais went into business serving emigrants traveling west. He established a road ranche some five miles east of Fort Laramie in 1853 and opened another on the South Platte at the "Old California Crossing" about twenty-five miles east of Julesburg in 1859. For his biography see Charles E. Hanson, Jr., "Geminien P. Beauvais" in LeRoy R. Hafen, ed., *The Mountain Men and the Fur Trade of the Far West* (Glendale, Calif.: Arthur H. Clark Co., 1969), 7:35-43.

[60] The 1860 census of the "Platte River Settlement" records S. H. Baker and H. B. Fales, as "traders." The Apr. 9, 1861, *Daily Telegraph* route guide says this ranche was also called Buckeye Ranche.

[61] Fremont's Orchard was on the South Platte River some one hundred miles southwest of Julesburg on the overland stagecoach line. A few miles beyond, the route divided, one branch going to Denver, the other skirting the east slope of the Rockies to rejoin the so-called Oregon Trail in today's Wyoming. Evidently the "orchard" appellation was derived from a grove of stunted cottonwood trees there that some travelers thought resembled an apple orchard. John C. Fremont traveled this way in 1843, but his journals make no mention of the orchard as such. Root and Connelley, *Overland Stage,* 224.

[62] P. T. Barnum had created an exhibit in his museum called "The Happy Family," which showed more than sixty animals and birds, "each of them being

the mortal enemy of every other" living together harmoniously. Phineas T. Barnum, *An Illustrated Catalogue and Guide Book to Barnum's American Museum* (New York: Wynkoop, Hallenbeck and Thomas, ca. 1860), 1.

[63] Lodgepole Creek runs roughly west to east in the southern Nebraska Panhandle, approximating the route of U.S. Highway 30, then turning southwesterly to cross the Colorado line and intersect the South Platte River west of Julesburg. Although Brown neglected to insert the estimated mileage here, he gave the mileage from the Julesburg telegraph station along Lodgepole Creek to the point where the line turned north to Mud Springs as about thirty-seven miles in his following entry of July 5.

[64] George H. Guy, a native of Otsego County, New York, came to Omaha in 1857 and freighted for five years before being employed by Creighton on the telegraph project. Andreas, *History of Nebraska,* 772.

[65] Jules Bene (Beni) was appointed to manage the Russell, Majors and Waddell stagecoach station here. As the story goes, Jules was dishonest and company division agent Joseph A. Slade determined to replace him, whereupon Jules shot and nearly killed Slade. When Slade recovered, he tracked Jules down, tied him to a post, shot him to pieces, and cut off his ears, one of which Slade used as a watch fob. Thrapp, *Encyclopedia,* 1:92.

[66] This Julesburg, which began as a station on the overland stagecoach route, was on the south side of the South Platte River and several miles southwest of the modern town, which is north of the river. Old Julesburg was sacked and burned by Indians in January and February 1865.

[67] Frank Root described one yell phonetically as "Ah-whooh-wah," but said only those who had actually heard it could really conceive of "the shrill and hideous-sounding noise." Another yell was "Yep, yep, yep," but "there was nothing specially hideous about it." Root and Connelley, *Overland Stage,* 88.

[68] This may be "Mr. Ellsworth" mentioned as the telegraph operator at Fort Kearny in the Nov. 2, 1860, issue of the *Huntsman's Echo.* In February 1865 Richard S. Ellsworth was an employee of the telegraph company at Mud Springs when the Indians attacked it, according to a C. F. Porter letter, Mar. 10, 1865, in Omaha *Nebraska Republican,* Mar. 31, 1865. Perhaps these references are to the same person. Randall A. and James J. Brown appear in the 1860 census of Omaha, both born in New York State, followed by the enumeration for C. H. Brown, though the latter was not living in the same household as his brothers. The Apr. 11, 1861, *Daily Telegraph* has an advertisement for J. J. and R. A. Brown, wholesale and retail dry goods merchants, Fourteenth and Douglas streets in Omaha.

[69] "Mr. Lo" was a common nickname applied to Indians and derived from a line in a poem by English poet Alexander Pope: "Lo! The poor Indian, whose untutored mind sees God in clouds, or hears him in the wind."

[70] The modern-day name of this tributary is Frenchman Creek, which joins the Republican River near Culbertson in Hitchcock County.

[71] This was the solar eclipse of May 26, 1854, visible in Brown's home state of New York, where he was then living. http://eclipse.gsfc.nasa.gov/eclipse.html, accessed May 12, 2010.

[72] Augustus and Peter Byram of Nebraska City succeeded the firm of Russell, Majors, and Waddell and "for several years they were the ranking freighters on the Plains." Lass, *From the Missouri to the Great Salt Lake,* 220-21.

[73] Missouri Governor Claiborne Jackson had refused the federal government's call for troops to help put down the rebellion and was working toward the goal of having Missouri secede from the Union. Louis S. Gerteis, *Civil War St. Louis* (Lawrence: University Press of Kansas, 2001), 93.

[74] The Nebraska Ranche, kept by Ackley and Forbes, as recorded in the *Daily Telegraph* route guide, Apr. 9, 1861.

[75] This was likely the so-called Great Comet of 1861, discovered by John Tebbutt of Australia. By late June it was visible in the northern hemisphere to the naked eye. www.phenomena.org.uk/cometof1861.htm accessed May 12, 2010.

[76] Hibbard had been in charge of constructing the telegraph line from Fort Kearny to Julesburg.

[77] Milan Hunt, age thirty-five, is listed as a "trader" in the 1860 census of Omaha.

[78] This was a Pony Express and stagecoach station about 3.5 miles east of today's Sidney. Here the trail and telegraph line crossed Lodgepole Creek and turned north to intersect the North Platte River near Court House Rock. Mattes, *Great Platte River Road,* 471.

[79] Mattes, ibid., 472, identifies this site as being about three miles south and one mile west of the village of Gurley. Ware, *Indian War of 1864,* 192, describes the "well" and indicated that it was dry.

[80] For the history of Mud Springs, see Paul Henderson, "The Importance of Mud Springs," *Nebraska History* 32 (June 1951): 108-19; and John D. McDermott, "'We had a Terribly Hard Time Letting Them Go': The Battles of Mud Springs and Rush Creek, February 1865," *Nebraska History* 77 (Summer 1996): 78-88. The Nebraska State Historical Society holds title to the station site.

[81] See Mattes, *Great Platte River Road,* chap. 11. The Nebraska State Historical Society also holds title to Court House-Jail Rocks.

[82] Wakeley, appointed associate justice of the Nebraska Supreme Court in 1857, had just been replaced in that position by an appointee of the Lincoln administration. In 1854 Wakeley married a Miss Comstock, thus "Young Comstock" mentioned subsequently by Brown must be Wakeley's brother-in-

law. Morton-Watkins, *Illustrated History of Nebraska,* 1:765-66.

[83] Modern maps show the small stream named Lawrence Fork as tributary to Pumpkin Creek, which flows generally west to east through central Morrill County, skirting Court House Rock on the south and then joining the North Platte River. Eugene Ware in 1864 had Lawrence Fork as the primary stream and Pumpkin Creek as the tributary. Lawrence is said to derive from a corruption of the name of a French fur trader named Lorren or Loran. Mattes, *Great Platte River Road,* 341-42; Ware, *Indian War of 1864,* 192.

[84] Alexander Benham was a division agent on the line between Julesburg and Denver, according to Root and Connelley, *Overland Stage,* 471. He is listed in the 1860 census for the "Platte River Settlement," as a twenty-six-year-old native of New York State.

[85] The distance from Mud Springs to the Platte River was probably about nine miles as the crow flies.

[86] Brown probably had no way of knowing the altitude when he was in the field, and intended to fill in the figure later. According to USGS topographic maps, the altitude at Mud Springs is slightly more than four thousand feet.

[87] Although this outfit was apparently jointly owned by John and David Hazard, Brown is never clear about whether both men were in the field. Since he mentioned David Hazard by name in connection with crossing the Platte near Julesburg, it is probably he who was in charge of the wagon train camped near Chimney Rock.

[88] Fewer than eight years would elapse before Brown's prediction of a transcontinental railroad would be realized.

[89] The outbreak of the Civil War forced the government to shift the transcontinental mail route northward to avoid having to go through Texas, part of the Confederacy. The relocation was to be effective July 1, 1861, and the coaches followed the Oregon-California trail via Fort Laramie in part because the Central Overland California and Pikes Peak Express had existing stations for the Pony Express along that route. At Julesburg, the mail route to Denver divided from the main trail. Lass, *From the Missouri to the Great Salt Lake,* 122-23.

[90] See note 67.

[91] Brown did not know at the time of his writing that the Pony Express was a financial failure. Those losses, along with other setbacks, had plunged Russell, Majors, and Waddell deeply into debt, much of it to Ben Holladay, who took over the COCPP in foreclosure in 1862. Lass, *From the Missouri to the Great Salt Lake,* 123.

[92] A paraphrase of a line from Longfellow's *Song of Hiawatha:* "And the evening sun descending set the clouds on fire with redness."

[93] Brown does not further identify Thompson. He must have been a teamster working for one of the freighting contractors Creighton had engaged for the construction of the telegraph line.

[94] Brown here follows the lead of the thousands of preceding fur traders, emigrants, missionaries, and soldiers whose diaries and journals recorded passing Chimney Rock, making it the most-mentioned landmark on the trail. Chimney Rock is a national historic site easily accessible to the modern traveler via Nebraska 92 near Bayard, Nebraska, and interpreted at a visitor center operated by the Nebraska State Historical Society.

[95] "Pete" was among many travelers who misjudged the distance of Chimney Rock from the trail and attempted the "short hike" to visit it.

[96] These "Black Hills" were the eastern foothills of the Laramie Range, not the region in today's South Dakota. Lass, *From the Missouri to the Great Salt Lake,* 40.

[97] There are many sources for the history of Fort Laramie, including Douglas C. McChristian, *Fort Laramie: Military Bastion of the High Plains* (Norman: Arthur H. Clark Co., 2009), which traces its years as a U.S. Army post, 1849-90. Brown is mistaken as to Fort Laramie's origin. It was established in 1834 by the firm of Sublette and Campbell as a fur trading post named Fort William. Fort Laramie National Historic Site is today a unit of the National Park Service.

[98] Brown gave the details of the accident that killed Wells in his introduction to the daily diary. The quote is part of the sentence, "here lies the body of Jeems Hambrick, who was accidentally shot on the bank of the Pecos River," which appeared in a book of humorous western tales by George Horatio Derby, writing under the pseudonym "John Phoenix." The book is entitled, *Phoenixiana: Or, Sketches and Burlesques* (N.Y.: D. Appleton & Co., 1856). Derby was a West Point graduate, Mexican War veteran, and U.S Army Topographical Engineer who served on the West Coast. www.sfmuseum.org/hist9/derby.html, accessed May 13, 2010.

[99] Horseshoe Creek flows through the Laramie Range to the north of Laramie Peak and enters the North Platte River southeast of present Glendo.

[100] These were two Pony Express and stagecoach stations, the latter located on Cottonwood Creek, also known as Bitter Cottonwood Creek. https://wyomingplaces.pbworks.com/cottonwood accessed May 13, 2010.

[101] He means Jules Ecoffey, who had been in the Fort Laramie vicinity for years. Ecoffey was probably working for Fort Laramie sutler Seth Ward, who also owned a ranche west of the fort. In 1860 Sir Richard Burton passed through "Ward's Station, alias the 'Central Star'." Burton, *The Look of the West, 1860: Across the Plains to California* (1862, rpt. Lincoln: University of Nebraska Press Bison Books), 113, 324. In the early 1870s Ecoffey formed a partnership with Adolph Cuny to operate a brothel and saloon north of Fort Laramie. McChristian, *Fort Laramie,* 197n32.

[102] This was Antoine Reynal, Jr., born in the French settlement of St. Charles, Missouri. See Charles E. Hanson, Jr., "Antoine Reynal, Jr," in Hafen, *Mountain Men and the Fur Trade,* 9:331-34. Sir Richard Burton stopped at Reynal's ranche in August 1860, where he ate breakfast consisting of "rusty bacon" and

antelope steak "cut off a corpse suspended for the benefit of flies outside." Burton, *The Look of the West,* 99-104.

[103] James Bordeaux, who had long been involved in the fur trade and Indian trade near Fort Laramie. The site of his wintering post near Chadron is home to the Museum of the Fur Trade. Bordeaux's biography by John D. McDermott is found in Hafen, *Mountain Men and the Fur Trade,* 5:65-80.

[104] William H. Hooper was returning from a trip to Washington, D.C. The Omaha Daily Telegraph of June 20, 1861, noted Hooper's arrival in Omaha on his way to the nation's capital. Hooper had once managed a Salt Lake City store owned in part by Ben Holladay. Lass, *From the Missouri to the Great Salt Lake,* 50.

[105] This area is probably the region known today as the Wildcat Hills.

[106] Common cathartics used at this time included mercurous chloride or Calomel.

[107] In modern terminology, bighorn sheep, which have recently been reintroduced in the Wildcat Hills and Pine Ridge areas of western Nebraska.

[108] He means the Fort Laramie Military Reservation.

[109] In 1864 telegraphers told Iowa cavalryman Eugene Ware about administering electrical shocks to Indians to make them dread tampering with the telegraph wire. Ware also heard a story about a group of Indians who cut down half a mile of wire near O'Fallon's Bluffs and were dragging it across the prairie when the wire was struck by lightning, knocking most of the men off their horses. Even if these events actually took place they, and the demonstration Brown describes, did not stop Indians from destroying miles of the telegraph line during their February 1865 raid on Julesburg. Ware, *Indian War of 1864,* 80-81.

[110] The previously identified ranche owned by Seth Ward west of Fort Laramie.

[111] This was Joseph A. "Jack" Slade, who supposedly killed Jules Bene of Julesburg fame and cut off his ears. Thrapp, *Encyclopedia,* 3:1317-18. See also note 65. When Richard Burton stopped at Horseshoe Station a year earlier, he said Slade was the station-keeper. Burton, *The Look of the West,* 113.

[112] A. W. Greely, "The Military-Telegraph Service," in Vol. 8 of *The Photographic History of the Civil War In Ten Volumes,* ed. Francis Trevelyan Miller (1911; rpt. New York: Thomas Yoseloff, 1957), 342.

[113] E. B. Long, *The Civil War Day-by-Day* (Garden City, N.Y.: Doubleday and Company, Inc., 1971), 725.

[114] Coe, *The Telegraph,* 52, 118; Jack Coggins, *Arms and Equipment of the Civil War* (Garden City, N.Y.: Doubleday and Co. Inc., 1962), 108.

[115] Coe, *The Telegraph,* 52, 61; Long, *Civil War Day-by-Day,* 725; Mike Wright, *What They Didn't Teach You About the Civil War* (Navato, Calif.: Presidio Press, 1996), 107; Robert D. Hoffsommer, "Telegraph," in *Historical*

Times Illustrated Encyclopedia of the Civil War, ed. Patricia L. Faust (New York: Harper Perennial, 1991), 745.

[116] Greely, "Military Telegraph Service," 348, 356.

[117] Doris Kearns Goodwin, *Team of Rivals: The Political Genius of Abraham Lincoln* (New York: Simon and Schuster, 2005), 507; Edward Rosewater left the U.S. Military Telegraph Corps in September 1863 to take a job in the Omaha office of Edward Creighton's Pacific Telegraph Company. He rose to become the manager of the Omaha office until January 1870, when he began his journalism career. Morton-Watkins, *Illustrated History of Nebraska* 1:745.

[118] See John D. McDermott, "Guardians of the Pacific Telegraph," *Annals of Wyoming* 83 (Winter 2011): 21-31.

[119] Coe, *The Telegraph,* 39, 98-99, 100-101.

[120] Ibid., 62; Clayton K S Chun, *The U.S. Army in the Plains Indian Wars,* 1865-91 (Oxford, U.K.: Osprey Publishing, 2004), 77.

[121] Coe, *The Telegraph,* 42; Chun, *U.S. Army,* 77-78.

[122] Rod Gragg, *The Old West Quiz and Fact Book* (New York: Harper and Row, 1986), 181; W. L. Holloway, *Wild Life on the Plains and Horrors of Indian Warfare* (St. Louis: Excelsior Publishing Company, 1891), 590, quoted in Richard E. Jensen, R Eli Paul, and John E. Carter, *Eyewitness at Wounded Knee* (Lincoln: University of Nebraska Press, 1991), 36.

[123] Coe, *The Telegraph,* 133-35.

[124] Ibid., 72-77, 169; Tomas Nonnenmacher, "History of the U.S. Telegraph Industry," at http://eh.net/encyclopedia/article/nonnenmacher.industry.telegraph.us, accessed February 24, 2011.

[125] Coe, *The Telegraph,* 123-26, 130, 134-35.

[126] Ibid., 47, 87-88; Standage, *The Victorian Internet,* 105-21; Nonnenmacher, "U.S. Telegraph Industry"; http://www.ftd.com/ftd-celebrates-100-years-ctg/product-ftd-history.

[127] Coe, *The Telegraph,* 121-22.

[128] Standage, *The Victorian Internet,* 133, 136-37.

[129] Coe, *The Telegraph,* 151-53.

[130] Ibid., 89.

[131] Ibid., 141-46, 151-53; Standage, *The Victorian Internet,* 196-97.

[132] Coe, *The Telegraph,* 151-52; Nonnenmacher, "U.S. Telegraph Industry."

[133] Coe, *The Telegraph,* 153; Nonnenmacher, "U.S. Telegraph industry"; http://en.wikipedia.org/wiki/Western-Union.

[134] Brownville *Nebraska Advertiser,* Aug. 30, 1860.

Bibliography

Unpublished Materials:

Charles Brown Manuscript. Western Union Telegraph Company Records. National Museum of American History, Smithsonian Institution.

Lamb, Squire. Papers. MS1561. Nebraska State Historical Society, Lincoln.

Population Schedule, Nebraska Territory. Eighth Census, 1860. National Archives Microfilm Publication M653, roll 665.

Population Schedule, Douglas County, Nebraska. Twelfth Census, 1900.

Books and Articles:

Andreas, A. T., comp. *History of Nebraska.* Chicago: Western Publishing Co., 1882.

Barnum, Phineas T. *An Illustrated Catalogue and Guide Book to Barnum's American Museum.* New York: Wynkoop, Hallenbeck, and Thomas, ca. 1860.

Burton, Richard. *The Look of the West, 1860: Across the Plains to California.* 1862. Lincoln: University of Nebraska Press Bison Books, rpt. n.d.

Chun, Clayton, K. S. *The U.S. Army in the Plains Indian Wars, 1865-91.* Oxford, U.K.: Osprey Publishing Co., 2004.

Coe, Lewis. *The Telegraph: A History of Morse's Invention and its Predecessors in the United States.* Jefferson, N. Car.: McFarland and Co., Inc., 1993.

Coggins, Jack. *Arms and Equipment of the Civil War.* Garden City, N.Y.: Doubleday and Co., 1962.

Gerteis, Louis S. *Civil War St. Louis.* Lawrence: University Press of Kansas, 2001.

Gilman, Musetta. *Pump on the Prairie: A Chronicle of a Road Ranch, 1859-1868.* Detroit: Harlo Press, 1975.

Goodwin, Doris Kearns. *Team of Rivals: The Political Genius of Abraham Lincoln.* New York: Simon and Schuster, 2005.

Gragg, Rod. *The Old West Quiz and Fact Book.* New York: Harper and Row, 1986.

Greely, A. W. "The Military-Telegraph Service." *The Photographic History of the Civil War in Ten Volumes.* Vol. 8. Ed. Francis Trevelyan Miller. 1911; New York: Thomas Yoselof, 1957.

Hanson, Charles E., Jr. "Geminien P. Beauvais." *The Mountain Men and the Fur Trade of the Far West.* Vol. 7. Ed. LeRoy R. Hafen. Glendale, Calif.: Arthur H. Clark Co., 1969.

______. "Antoine Reynal, Jr." *The Mountain Men and the Fur Trade of the Far West.* Vol. 9. Ed. LeRoy R. Hafen. Glendale, Calif.: Arthur H. Clark Co., 1972.

Henderson, Paul. "The Importance of Mud Springs." *Nebraska History* 32 (June 1951).

Hoffsommer, Robert D. "Telegraph." *Historical Times Illustrated Encyclopedia of the Civil War.* Ed. Patricia L. Faust. New York: Harper Perennial, 1991.

Jensen, Richard E., R. Eli Paul, and John E. Carter. *Eyewitness at Wounded Knee.* Lincoln: University of Nebraska Press, 1991.

Johnson, Harrison. *Johnson's History of Nebraska.* Omaha: Henry Gibson, 1880.

Lass, William E. *From the Missouri to the Great Salt Lake: An Account of Overland Freighting.* Lincoln: Nebraska State Historical Society, 1972.

Long. E. B. *The Civil War Day-by-Day.* Garden City, N.Y.: Doubleday and Co., 1971.

McChristian, Douglas C. *Fort Laramie: Military Bastion of the High Plains.* Norman, Okla.: Arthur H. Clark Co., 2009.

McDermott, John D. "James Bordeaux." *The Mountain Men and the Fur Trade of the Far West.* Vol. 5. Ed. LeRoy R. Hafen. Glendale, Calif.: Arthur H. Clark Co., 1968.

______. "'We had a Terribly Hard Time Letting them Go': The Battles of Mud Springs and Rush Creek, February 1865." *Nebraska History* 77 (Summer 1996).

Mattes, Merrill J. *The Great Platte River Road: The Covered Wagon Mainline via Fort Kearny to Fort Laramie.* Lincoln: Nebraska State Historical Society, 1969.

Mihelich, Dennis N. *The History of Creighton University: 1878-2003.* Omaha: Creighton University Press, 2006.

Morton, J. Sterling, succeeded by Albert Watkins. *Illustrated History of Nebraska.* Vol. 1. Lincoln: Jacob North and Co., 1905.

Nonnenmacher, Tomas. "History of the U.S. Telegraph Industry." http://eh.net/encyclopeida.

Parks, Douglas R. "Pawnee." *Handbook of North American Indians,* Vol. 13, Pt. 1: Plains. Ed. Raymond J. DeMallie. Washington, D. C.: Smithsonian Institution, 2001.

Pfeiffer, Benjamin. "The Role of Joseph E. Johnson and his Pioneer Newspapers in the Development of Territorial Nebraska." *Nebraska History* 40 (June 1959).

Phoenix, John [George H. Derby]. *Phoenixiana: Or, Sketches and Burlesques.* New York: D. Appleton and Co., 1856.

Potter, James E. "Firearms on the Overland Trails." *Overland Journal* 9 (1991).

______ "The Legend of Rawhide Revisited." *Nebraska History* 85 (Fall 2004).

Root, Frank A. and William E. Connelley. *The Overland Stage to California.* Topeka, Kan.: The Authors, 1901.

Standage, Tom. *The Victorian Internet.* New York: Walker and Co., 1998.

Streissguth, Thomas. *Communications: Sending the Message.* Minneapolis: The Oliver Press, 1997.

Sandweiss, Lee Ann, ed. *Seeking St. Louis: Voices from a River City, 1670-2000.* St. Louis: Missouri Historical Society Press, 2000.

Savage, James W. and John T. Bell. *History of the City of Omaha,*

Nebraska. New York: Munsell and Co., 1894.

Stolley, William. "History of the First Settlement of Hall County, Nebraska." *Nebraska History,* Special Issue (April 1946).

Thompson, Robert Luther. *Wiring a Continent.* Princeton, N.J.: Princeton University Press, 1947.

Thrapp, Dan. *Encyclopedia of Frontier Biography.* 3 vol. Lincoln: University of Nebraska Press Bison Books, 1991.

Wakeley, Arthur C. *Omaha: The Gate City and Douglas County, Nebraska.* Chicago: The S. J. Clarke Publishing Co., 1917.

The War of the Rebellion: A Compilation of the Official Records of the Union and Confederate Armies. 128 vol. Washington, D. C. : GPO, 1880-1901.

Ware, Eugene F. *The Indian War of 1864.* Ed. Clyde C. Walton. Lincoln: University of Nebraska Press, 1960.

Wright, Mike. *What They Didn't Teach You About the Civil War.* Navato, Calif.: Presidio Press, 1996.

Internet Sources:

http://www.ieeeghn.org/index.php/transcontinental_telegraph_line_%28US%29

http://www.phenomena.org.uk/cometof1861.htm

http://ww.sfmuseum.org/hist9/derby.html

https://wyomingplaces.pbworks.com/cottonwood

http://eclipse.gsfc.nasa.gov/eclipse.html

http://eh.net/encyclopedia/article/nonnenmacher.industry.telegraph.us

http://www.ftd.com/ftd-celebrates-100-years-ctg/product-ftd-history

http://en.wikipedia.org/wiki/Western-Union

Newspapers:

Council Bluffs (Ia.) *Nonpareil*

Brownville, *Nebraska Advertiser*

Huntsman's Echo (Wood River, N.T.)

Omaha Bee

Omaha Daily Herald

Omaha, *Daily Nebraskian*

Omaha Daily Republican

Omaha, *Daily Telegraph*

Omaha World-Herald (Eve.)

Omaha World-Herald (Morn.)

Index

References to illustrations are in italic type

Index

Index

Index

Acknowledgements

The authors thank Dr. Michael Tate, University of Nebraska at Omaha, for his review of the manuscript and suggestions.

From Dennis Mihelich: I want to thank Alison Oswald, Archivist, National Museum of American History, Smithsonian Institution, for providing a copy of the Brown Diary. I also thank Libby Krecek, registrar, and Gary Rosenberg, research specialist, at the Douglas County Historical Society for their help in my research on Brown and Edward Creighton. I thank Jim Potter for his careful reading of the prologue and epilogue and his many valuable suggestions. I dedicate my work in this volume to my wife, Joanne, and to my daughter Heidi.

Dennis N. Mihelich is a retired historian, past president of the Nebraska State Historical Society, and the author of a score of articles on Nebraska history.

James E. Potter is Senior Research Historian with the Nebraska State Historical Society and Associate Editor of *Nebraska History* magazine. He is the author of *Standing Firmly by the Flag: Nebraska Territory, the Civil War, and the Coming of Statehood, 1861-1867* (forthcoming from University of Nebraska Press).